GW00984607
To Tony
from
Look after your microbes Tony!
very best wishes
Laura Gowers

Listen to Your Microbes

First published in 2023 by
Liberties Press, Dublin, Ireland
libertiespress.com

Distributed in the UK by Casemate UK
casematepublishing.co.uk

Distributed in the US and Canada by Casemate IPM
casematepublishers.com

ISBN (print): 978-1-912589-39-5
ISBN (ebook): 978-1-912589-40-1

2 4 6 8 10 9 7 5 3 1
This title is available in the National Library of Ireland, and a CIP record of this title is available from the British Library.
All text by Fergus Shanahan
All images and design by Laura Gowers
Cover design by Baker incorporating original images by Laura Gowers
Printed in China by Jinhao Printing
The lines from William Carlos Williams's poem "The Orchestra" quoted on page 115
originally appeared in *The Desert Music* (New Directions, 1952).

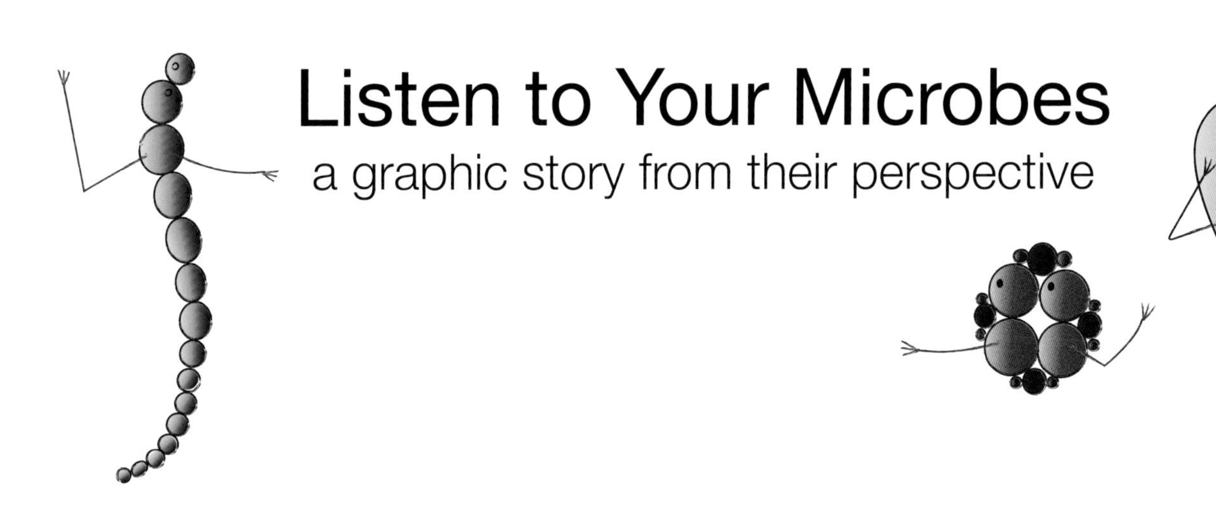

Listen to Your Microbes

a graphic story from their perspective

Fergus Shanahan and Laura Gowers

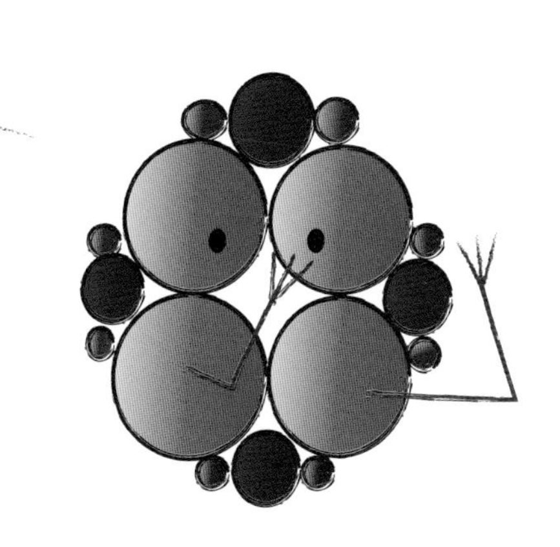
LISTEN!

DEDICATED TO THOSE WHO LISTEN
RATHER THAN WAIT TO SPEAK!

contents

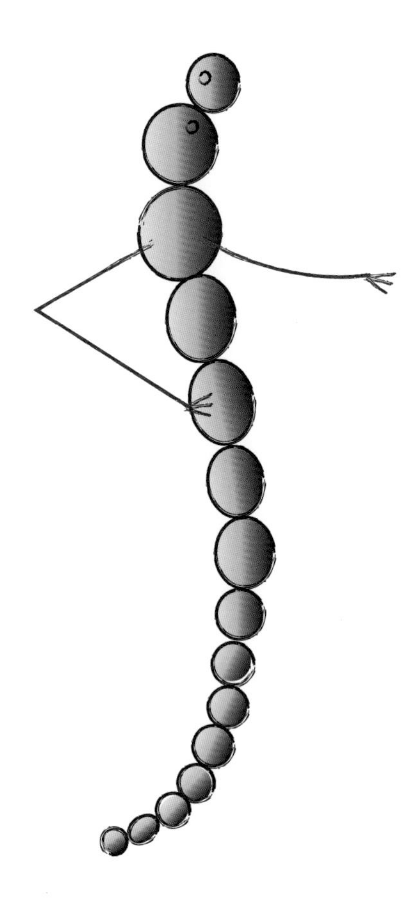

preface

This is a story for everyone, young and not young. The story of the microbes living in and on the human body – collectively known as the microbiome – is important for understanding personal and public health.

The authors are inspired by images drawn more than three hundred years ago by the first person to discover tiny creatures living in and on the human body: Antonie van Leeuwenhoek. Like that great discoverer, we are intrigued and charmed by the diversity, beauty and ingenuity of these wonders of nature. When a pandemic virus can bring our world to a standstill, bringing with it many lessons about human behaviour, we are reminded that we live in a microbial world. We hope that images of personal microbes will feed our readers' curiosity and their respect for the microbes we depend on.

We have included QR codes linked to further information for anyone wishing to pursue some of the issues in greater detail.

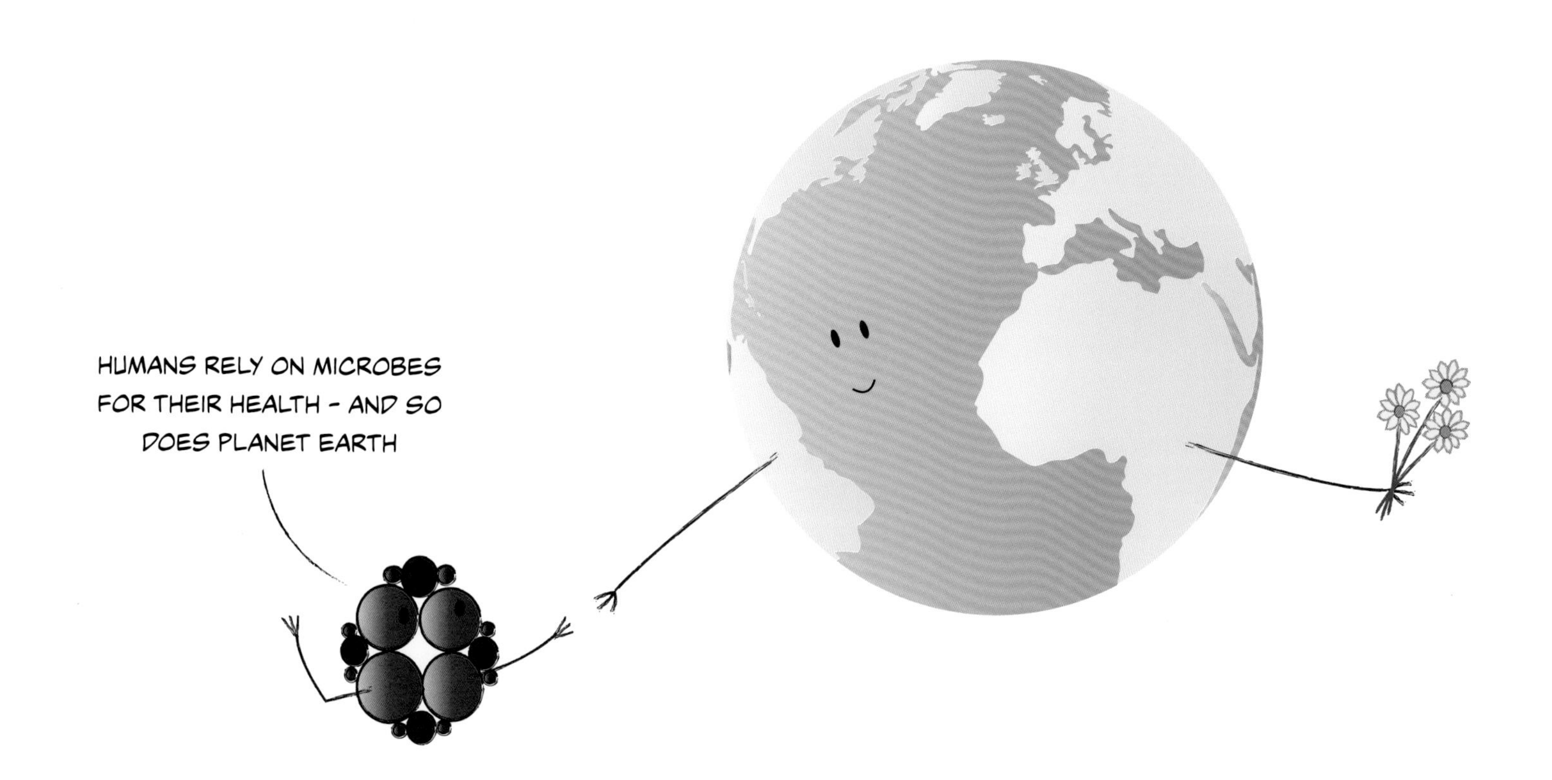
HUMANS RELY ON MICROBES FOR THEIR HEALTH - AND SO DOES PLANET EARTH

evolving together

Microbes (also known as micro-organisms) were on the planet long before humans. Microbes made the planet habitable for humans. Humans would not exist if ancient microbes had not mutated such that they could split water and produce oxygen. Microbes in the ocean also remove methane that seeps through from Earth's interior. Without microbes there would be no food, no plants and therefore no animals. Plants need specialised microbes in the soil to transform nitrogen (N_2) from the air to ammonia (NH_3) which they can use to produce chlorophyll, proteins, nucleic acids and the energy molecule ATP. Microbes also condition the soil by breaking down dead plant and animal material.

Humans also rely on personal microbes for their health. If microbes could speak, they would have much to teach us . . . if we would listen.

personal microbes

Evolution ensured that the microbes occupying the human body have been selected to promote health. Everyone has personal microbes – bacteria, archaea, viruses and fungi – that live in and on the human body. This community of microbes is known as the **microbiome**. Humans can't survive without personal microbes because they help digest food, train the immune system and support all body systems. Most of our personal microbes live in the gut, which is otherwise known as the digestive tract, from the mouth to the anus.

living in a microbial world

Microbes not only make the planet habitable, personal microbes also help humans to resist infections and survive in a world full of microbes.

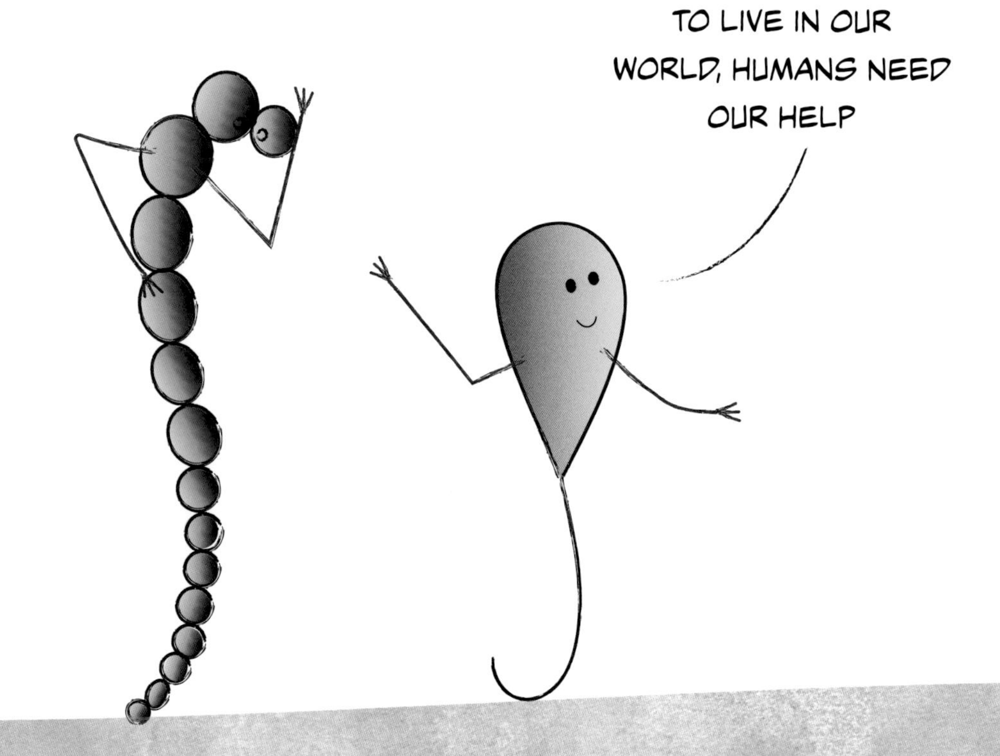

YEAH, OTHERWISE THEY WOULD NEED TO LIVE IN A BUBBLE
MICROBE FREE SPACE FOR HUMANS!

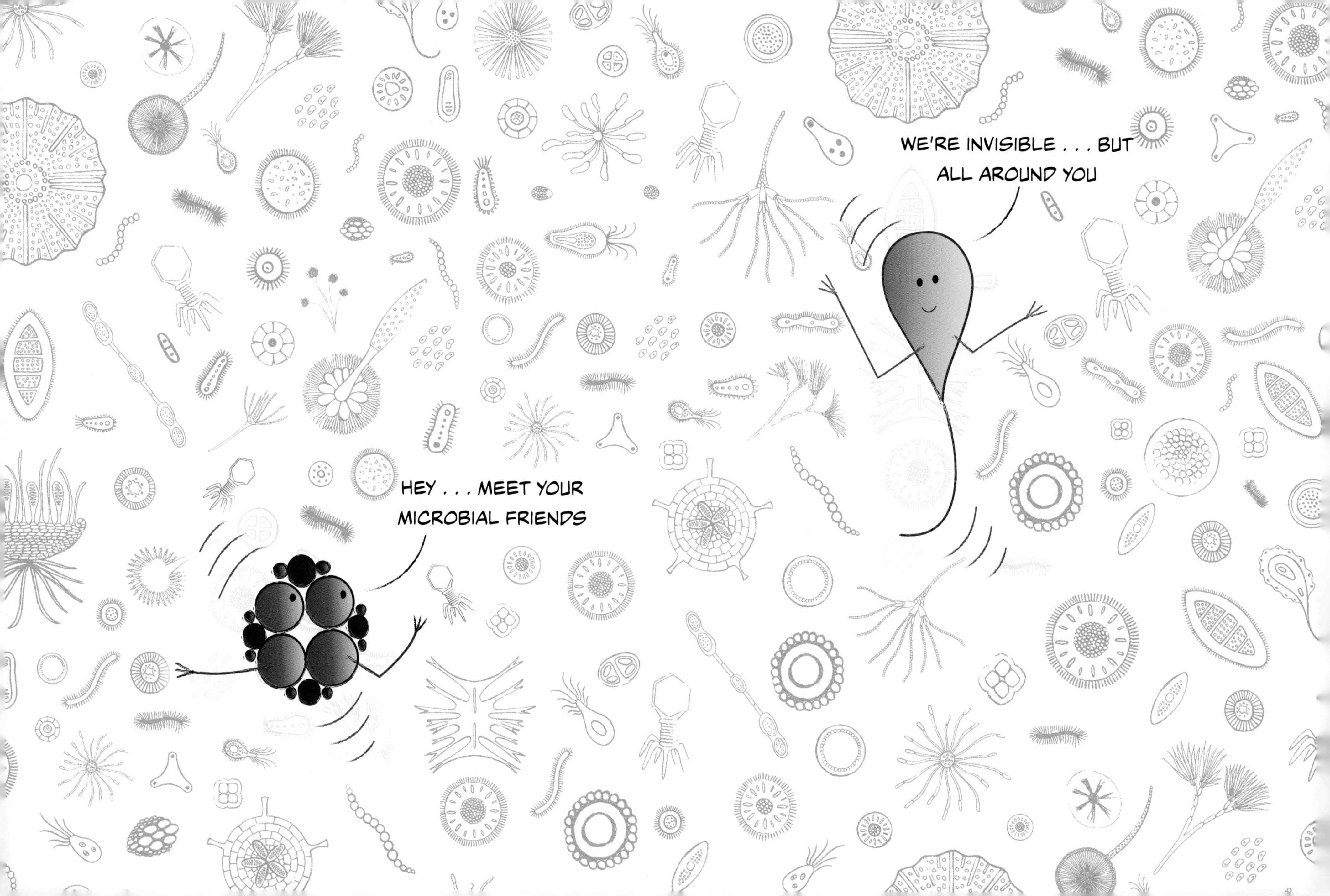
WE'RE INVISIBLE . . . BUT ALL AROUND YOU
HEY . . . MEET YOUR MICROBIAL FRIENDS

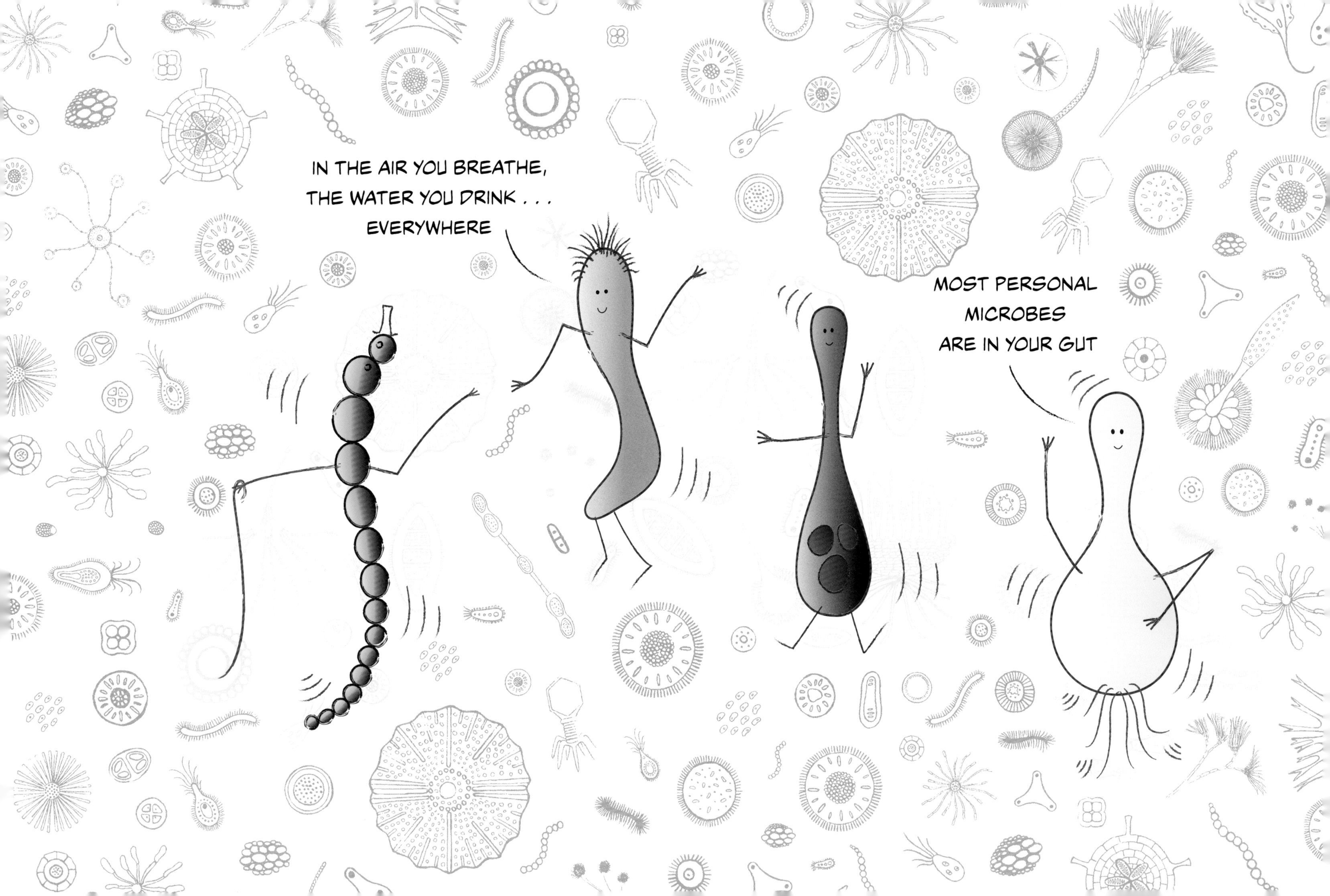
IN THE AIR YOU BREATHE,
THE WATER YOU DRINK . . .
EVERYWHERE
MOST PERSONAL
MICROBES
ARE IN YOUR GUT

I DIGEST THE
FOODS THAT HUMANS
CAN'T DIGEST

I TRAIN THE IMMUNE
SYSTEM TO IDENTIFY AND
RESPOND TO DANGER

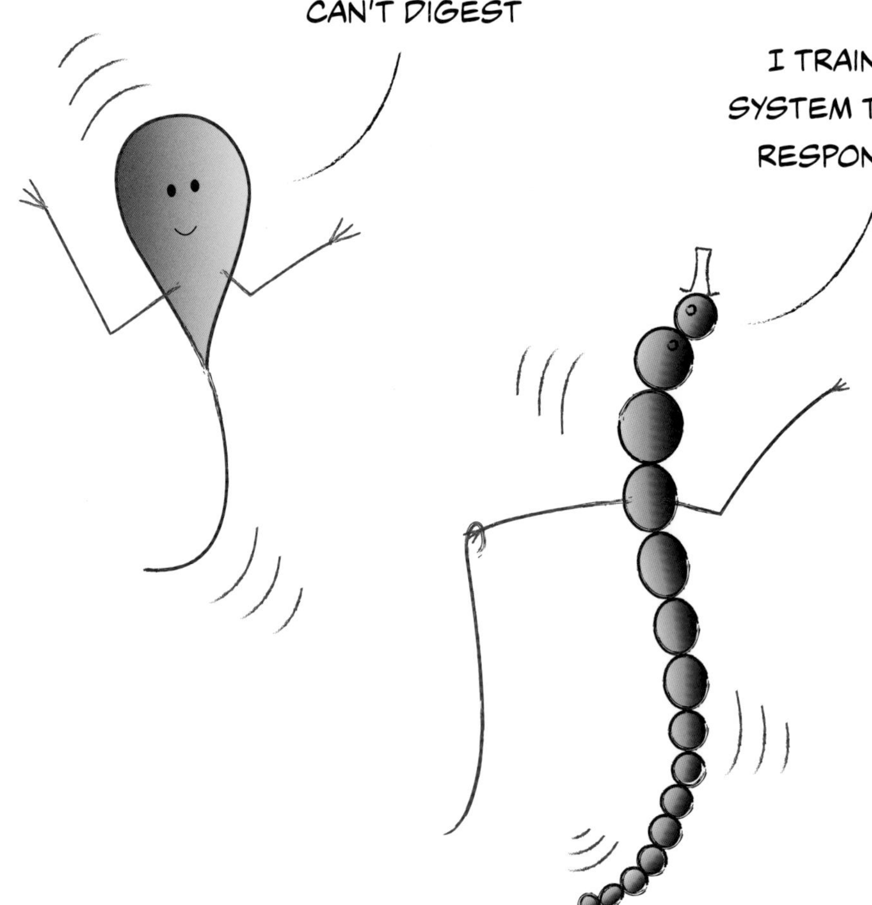

I CONTROL THE
NUMBERS OF MICROBES,
TO AVOID OVERCROWDING

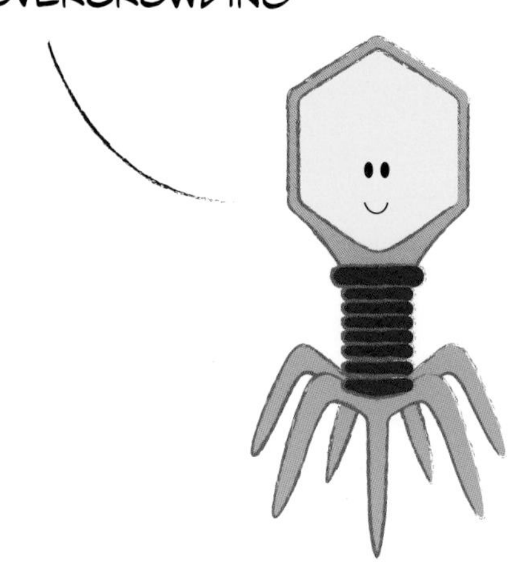

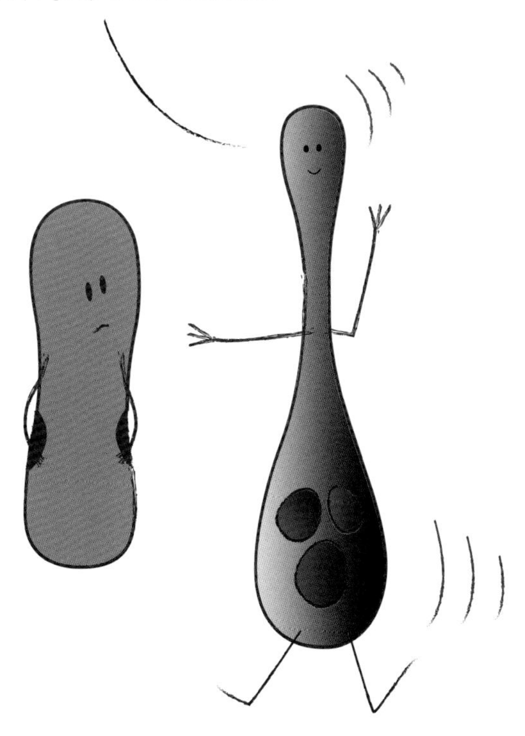

an organ within an organ

The collective metabolic activity of microbes in the gut is so central to how humans function that it could be considered as an additional vital organ. Gut microbes digest dietary fibres, protect against infections, train the immune system, regulate metabolism and contribute to brain development and behaviour. In addition, chemical signals from these same microbes keep the brain and the immune system aware of what's happening in the gut so that they can respond quickly to infections or when anything goes wrong.

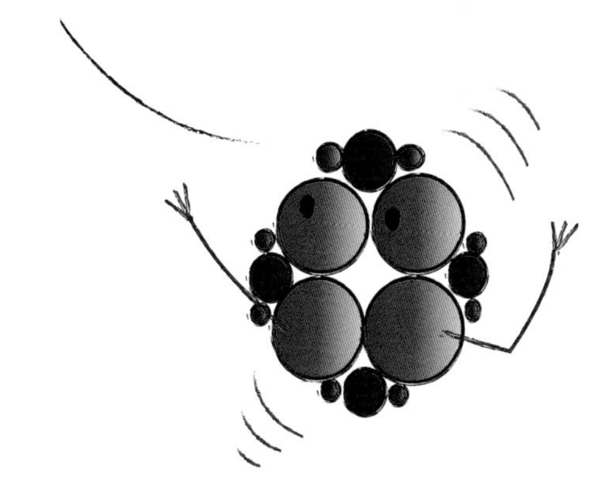

The brain-gut axis is a two-way communication system linking the brain with the gut. Signals from the gut travel to the brain, along nerves and through the blood stream, to keep the brain informed of what's happening in the gut. Many of these signals are metabolites produced by gut microbes when they digest dietary nutrients. The brain responds by signalling along the same routes to regulate the contents of the gut, the secretion of digestive juices and the gut's immune defences.

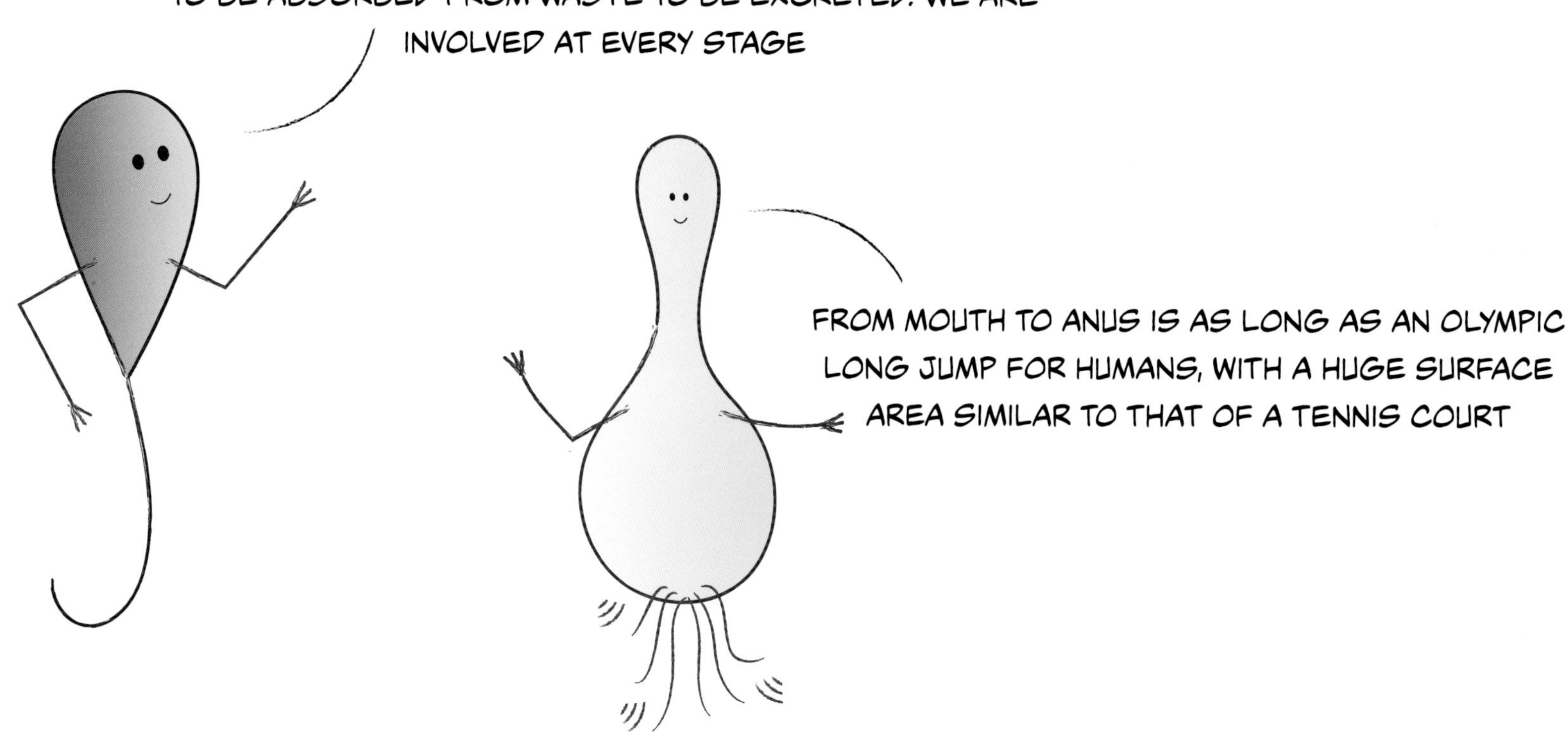

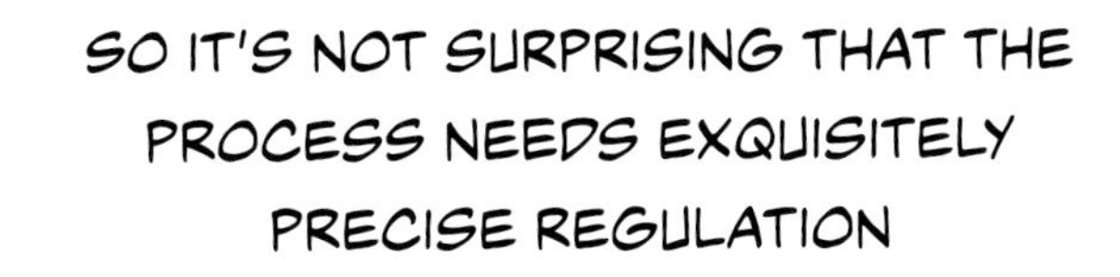
SO IT'S NOT SURPRISING THAT THE PROCESS NEEDS EXQUISITELY PRECISE REGULATION

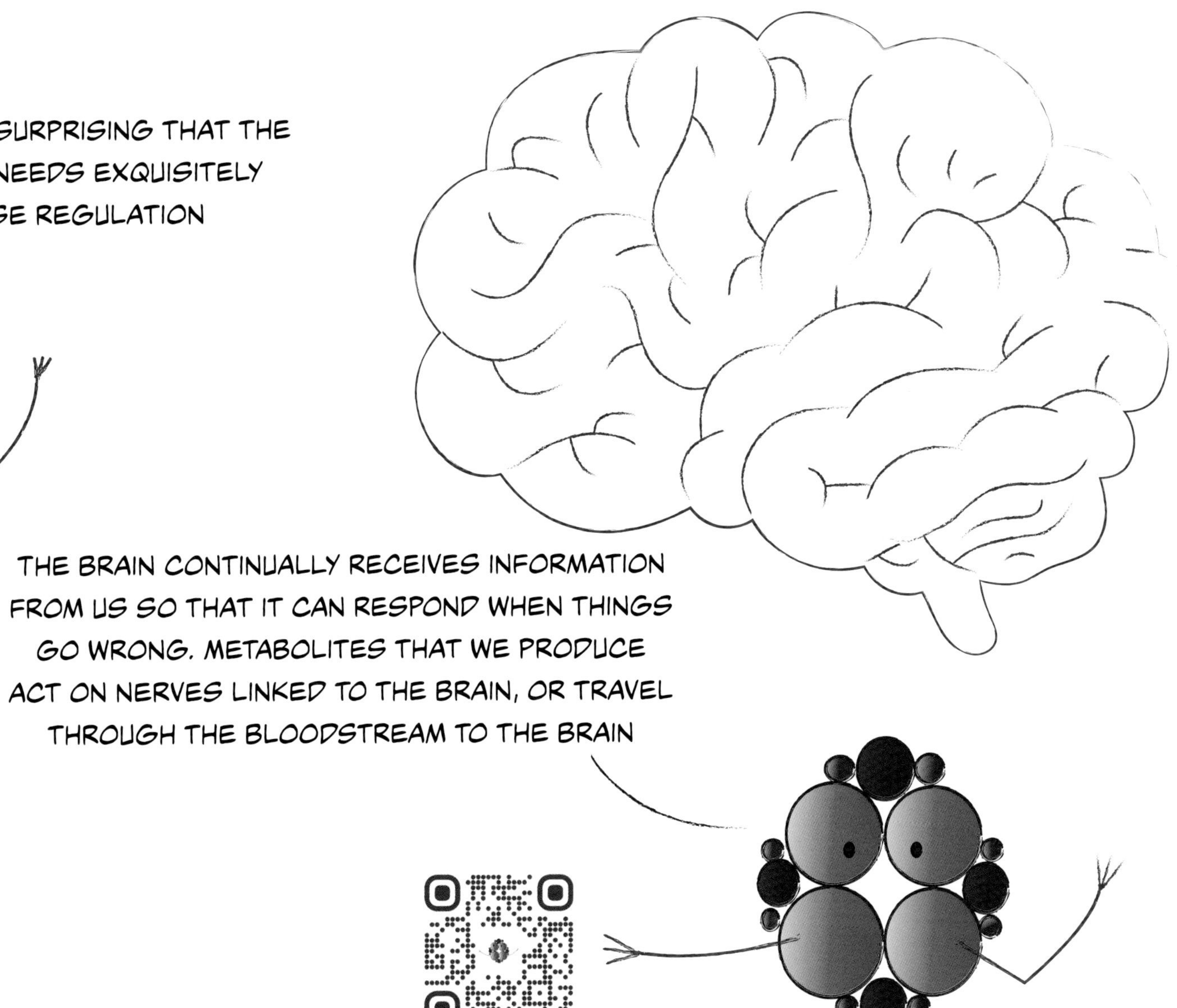
THE BRAIN CONTINUALLY RECEIVES INFORMATION FROM US SO THAT IT CAN RESPOND WHEN THINGS GO WRONG. METABOLITES THAT WE PRODUCE ACT ON NERVES LINKED TO THE BRAIN, OR TRAVEL THROUGH THE BLOODSTREAM TO THE BRAIN

YOUR MICROBIOME
IS AS UNIQUE AS YOUR
FINGERPRINT

BUT YOUR FINGERPRINT
ONLY TELLS WHO YOU ARE

HOWEVER, YOUR MICROBIOME
REVEALS HOW YOU LIVE AND
HOW HEALTHY YOU ARE!

2 discovering each other

What must it have been like for the first person to see microbes – tiny creatures living in and on and all around the human body – something that no man or woman had ever seen before, or even contemplated?

This is how it happened.

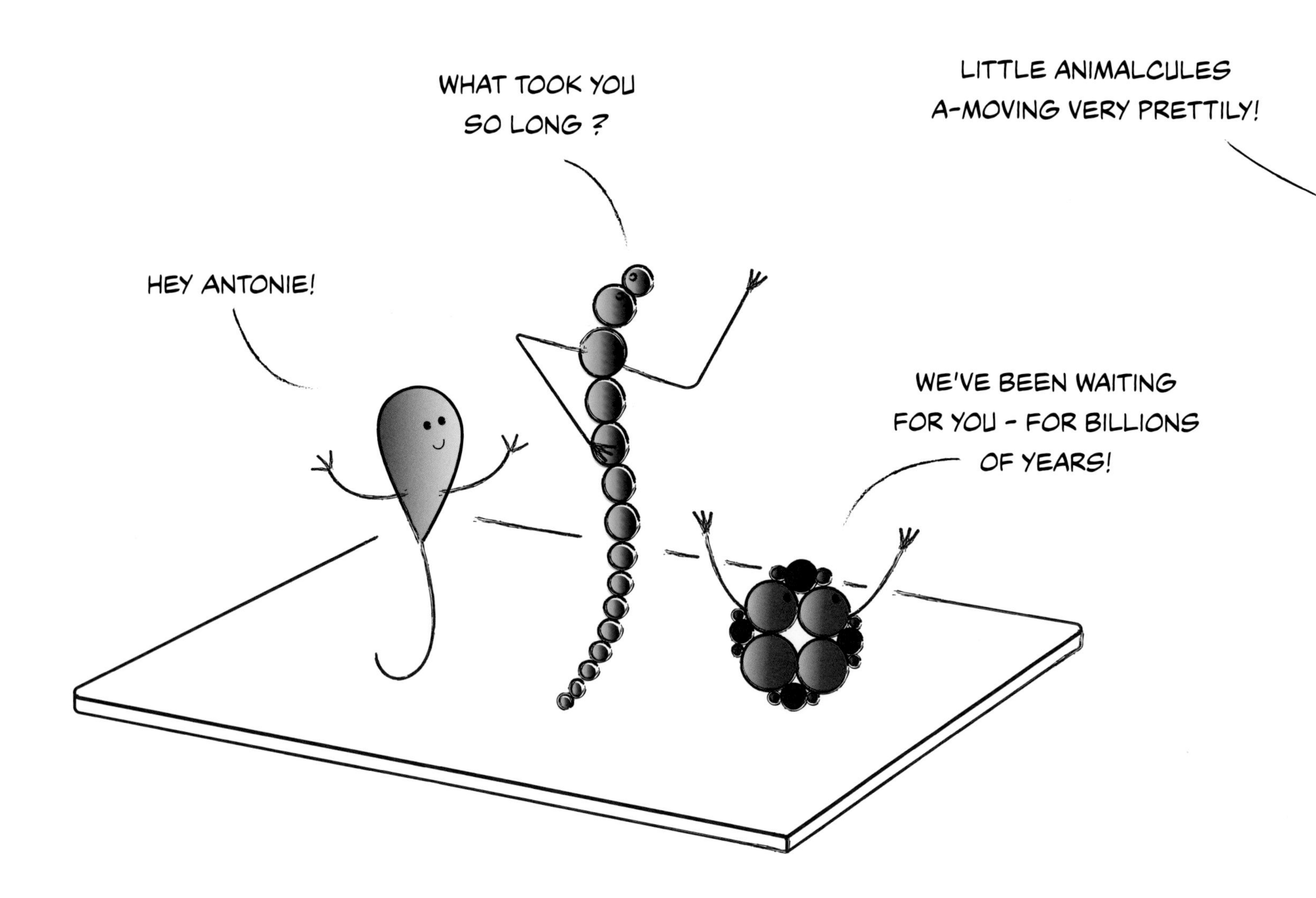
LITTLE ANIMALCULES A-MOVING VERY PRETTILY!
HEY ANTONIE!
WHAT TOOK YOU SO LONG ?
WE'VE BEEN WAITING FOR YOU - FOR BILLIONS OF YEARS!

local hero

Antonie van Leeuwenhoek (1632-1723)
was the first person to see his own microbes when he turned his home-made microscope on his own faeces and dental scrapings. We know from his writings that far from being repulsed by what he saw, he was enchanted.

and then the world forgot . . .

The importance of Antonie's findings was realised hundreds of years later. Today, we know that most microbes are beneficial; harmfulness (also called pathogenicity) is rare and usually due to a sort of biological accident when microbes are in the wrong place or are misinterpreted by the host.

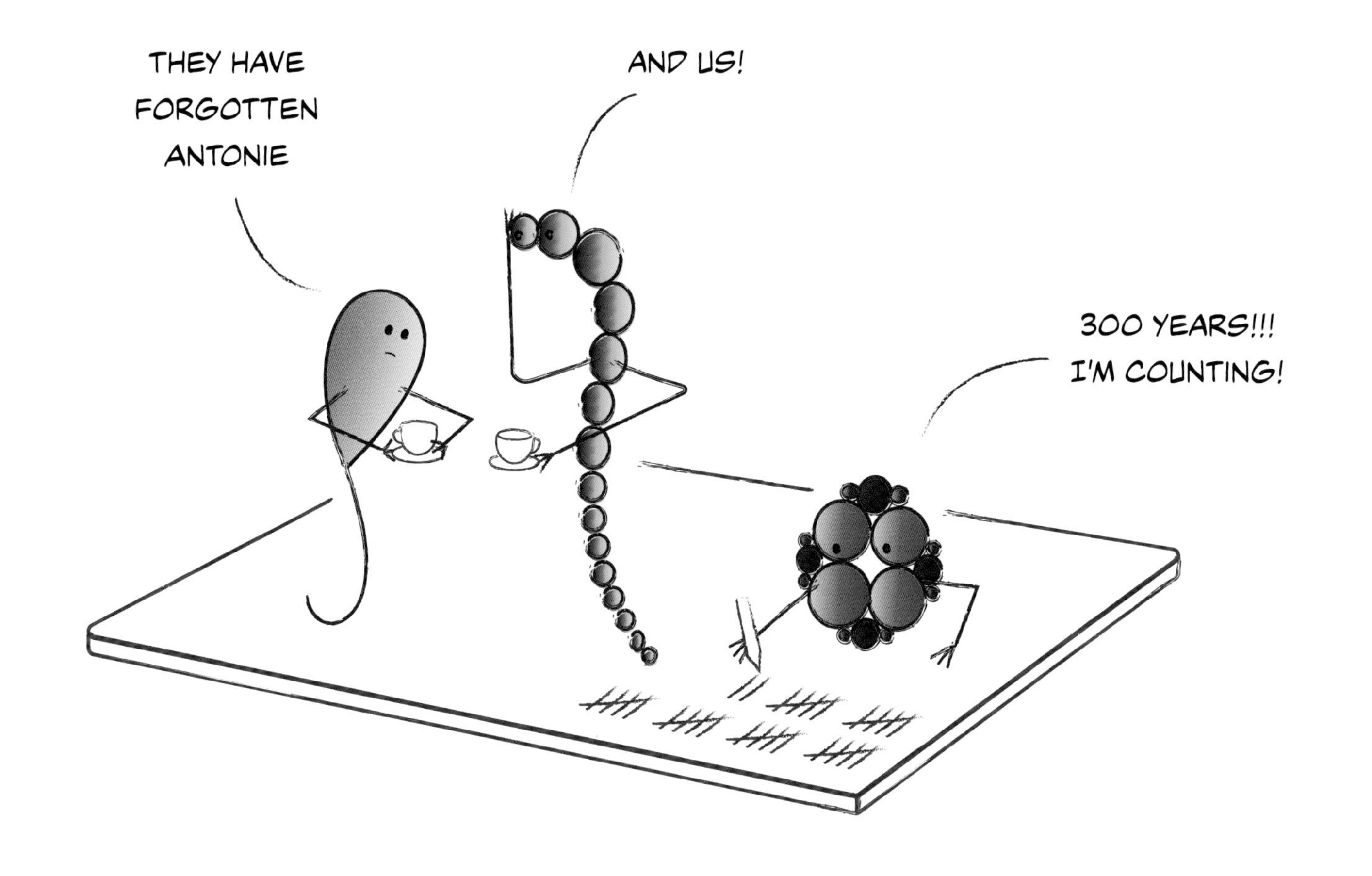
THEY HAVE FORGOTTEN ANTONIE
AND US!
300 YEARS!!! I'M COUNTING!

Charles Darwin (1809-1882) is credited with the discovery of evolution by natural selection, but only described species that he could see, whereas most life on the planet is microscopic. The story of evolution is incomplete if microbes are not included.

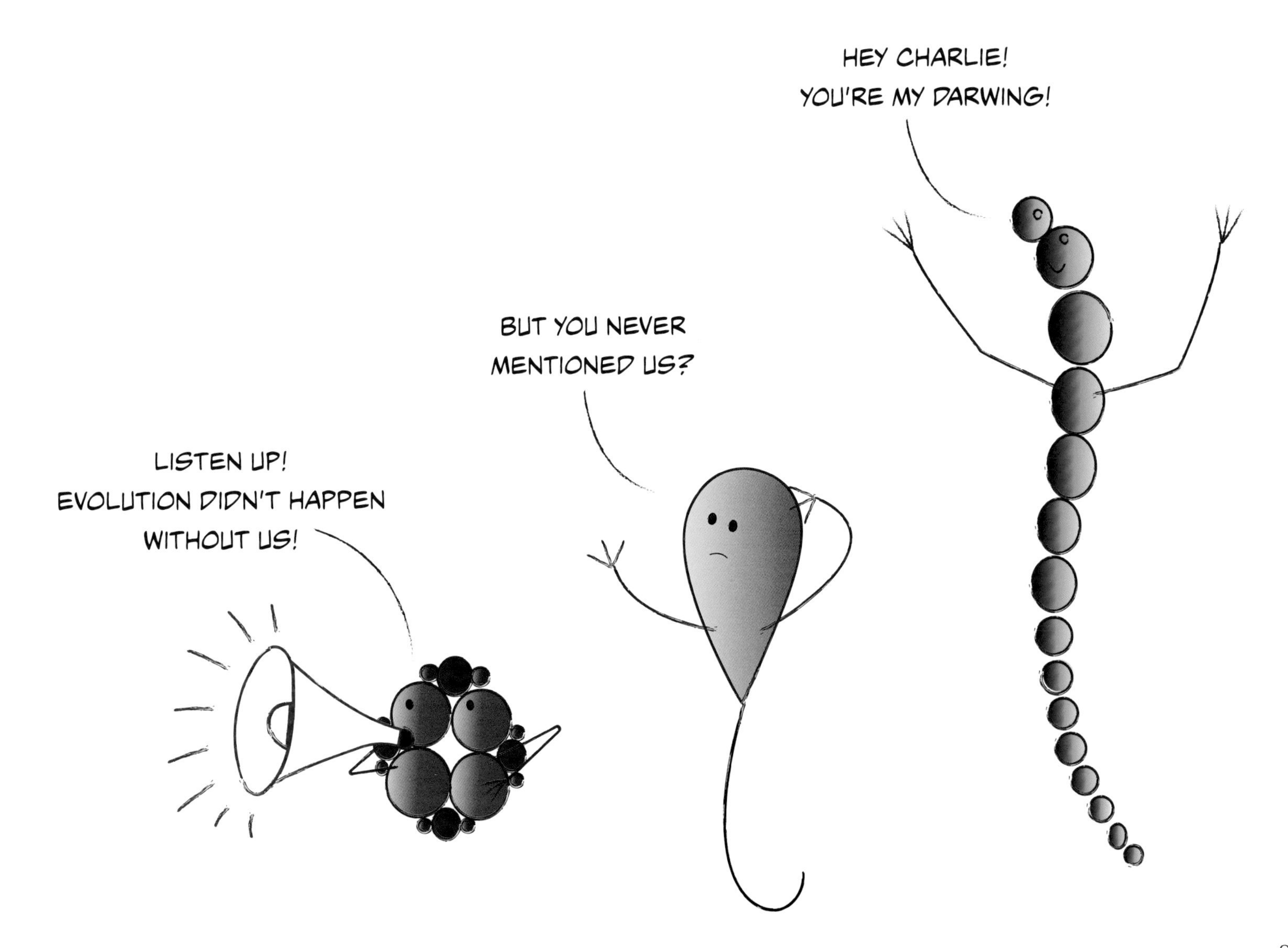
LISTEN UP!
EVOLUTION DIDN'T HAPPEN
WITHOUT US!
BUT YOU NEVER
MENTIONED US?
HEY CHARLIE!
YOU'RE MY DARWING!

Furthermore, **biodiversity** on the planet is mainly microscopic – out of sight. Loss of diversity, and so microbial extinctions, may go unrecorded, particularly when human activity changes abruptly. The same happens to the personal microbes of humans.

Changes in human lifestyle that accompany industrialisation have a major impact on the composition of microbes in the human gut. The changes that are associated with industrialisation and socio-economic development include: sedentary occupation and reduced physical activity, more time spent indoors, urban rather than rural development, greater consumption of antibiotics and other medications, smaller family size, and a low-fibre diet enriched with fat and refined sugars.

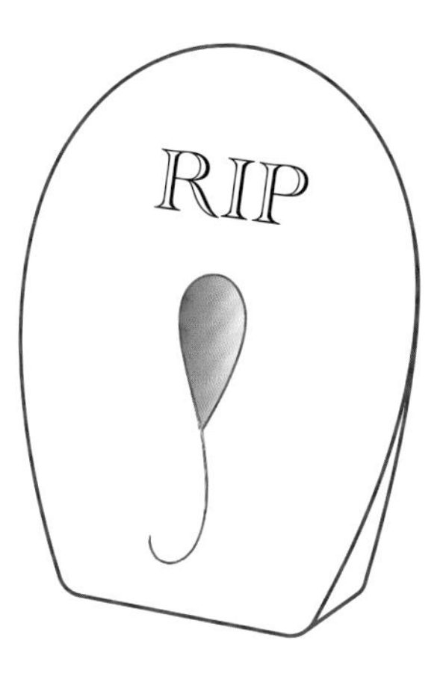

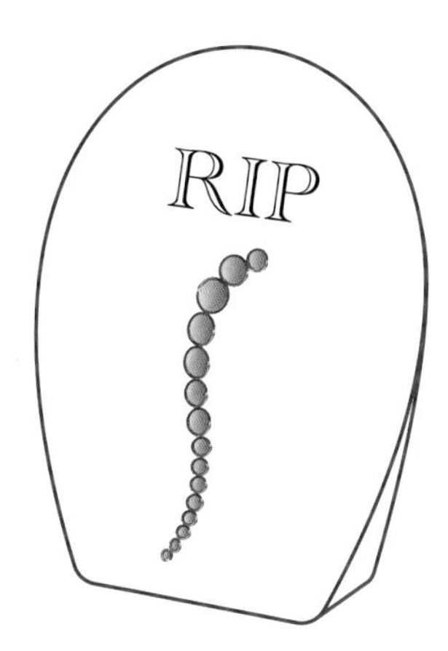

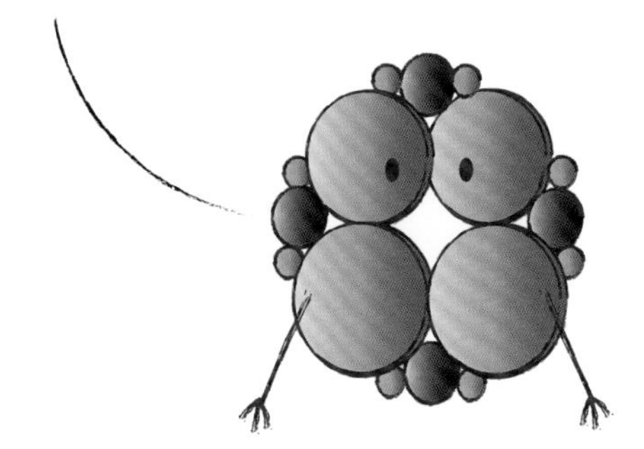
THIS IS WHAT HUMANS CALL 'MODERN' . . .
THEY CAN'T SEE HOW IT AFFECTS THEIR
MICROBES . . . FEWER ABLE TO DIGEST FIBRE,
MORE WITH ANTIBIOTIC-RESISTANCE GENES

OH DEAR!

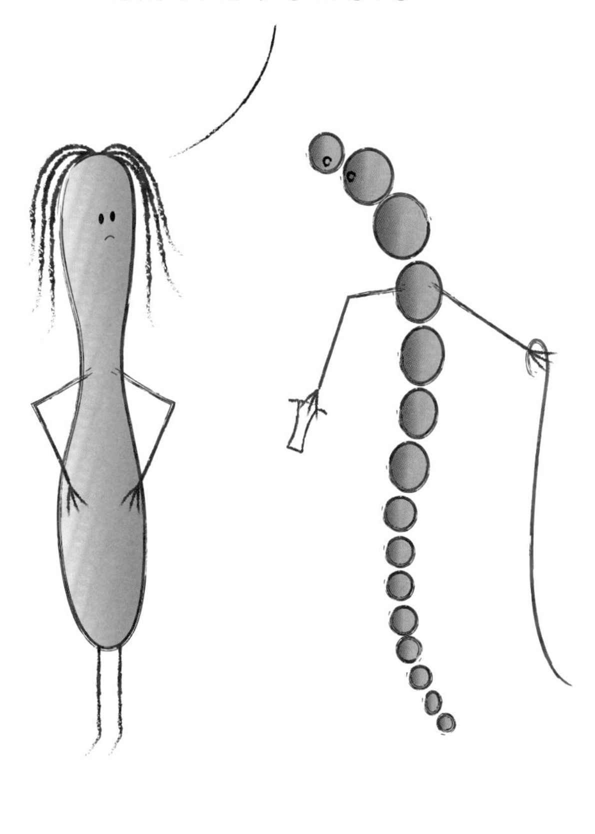
IT'S NOT SURPRISING
THAT HUMANS GET MORE
CHRONIC DISEASES

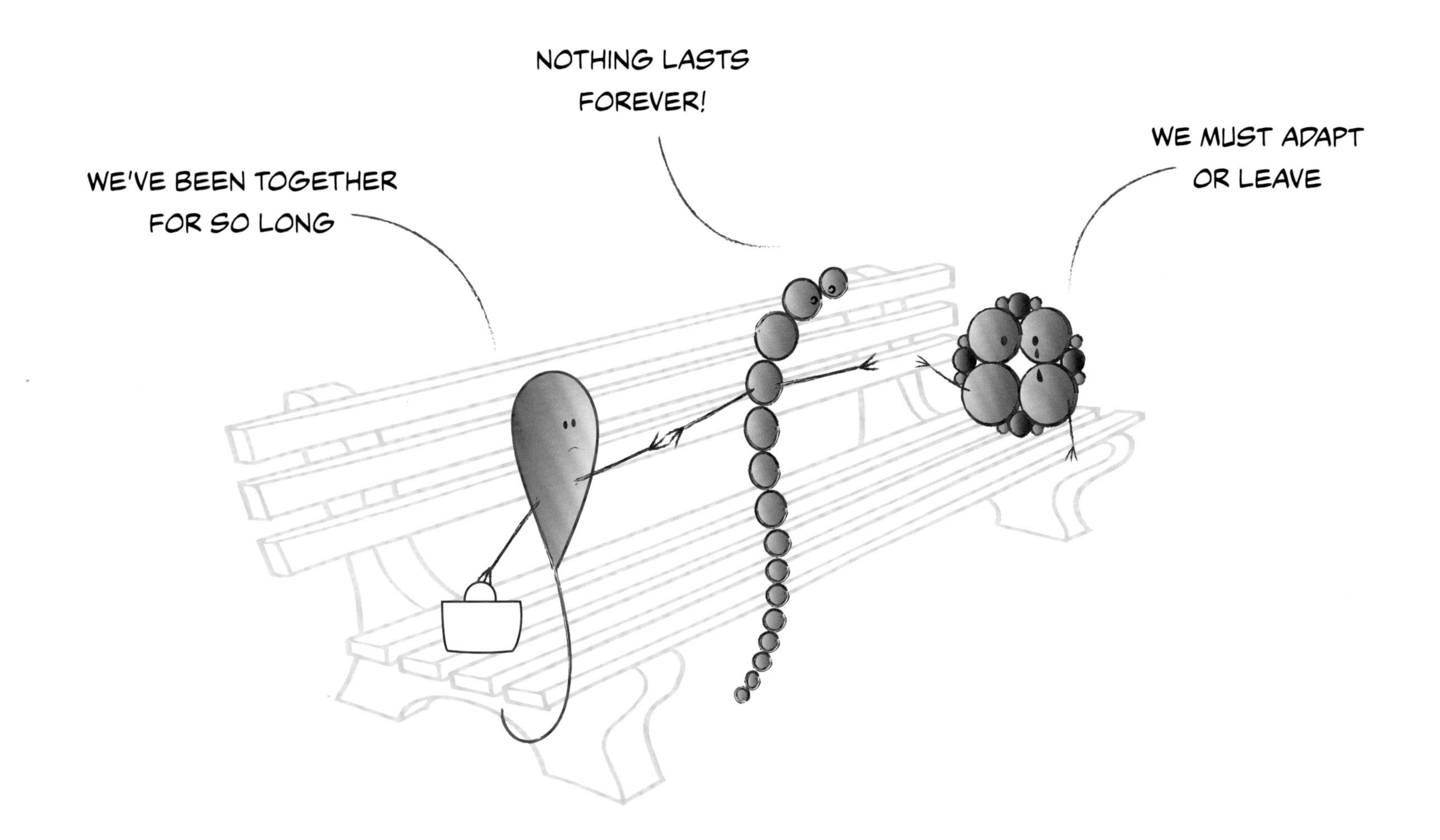
WE'VE BEEN TOGETHER FOR SO LONG
NOTHING LASTS FOREVER!
WE MUST ADAPT OR LEAVE

old friends

The loss of our ancestral microbes such as *Helicobacter pylori*, *Oxalobacter* and *Bifidobacterium spp* due to modernisation and Western diets has been associated with an increased risk of chronic inflammatory and metabolic diseases.

humans have several microbiomes

Different groups of microbes are adapted to different body-sites and even adapt to different niches within the same body-site.

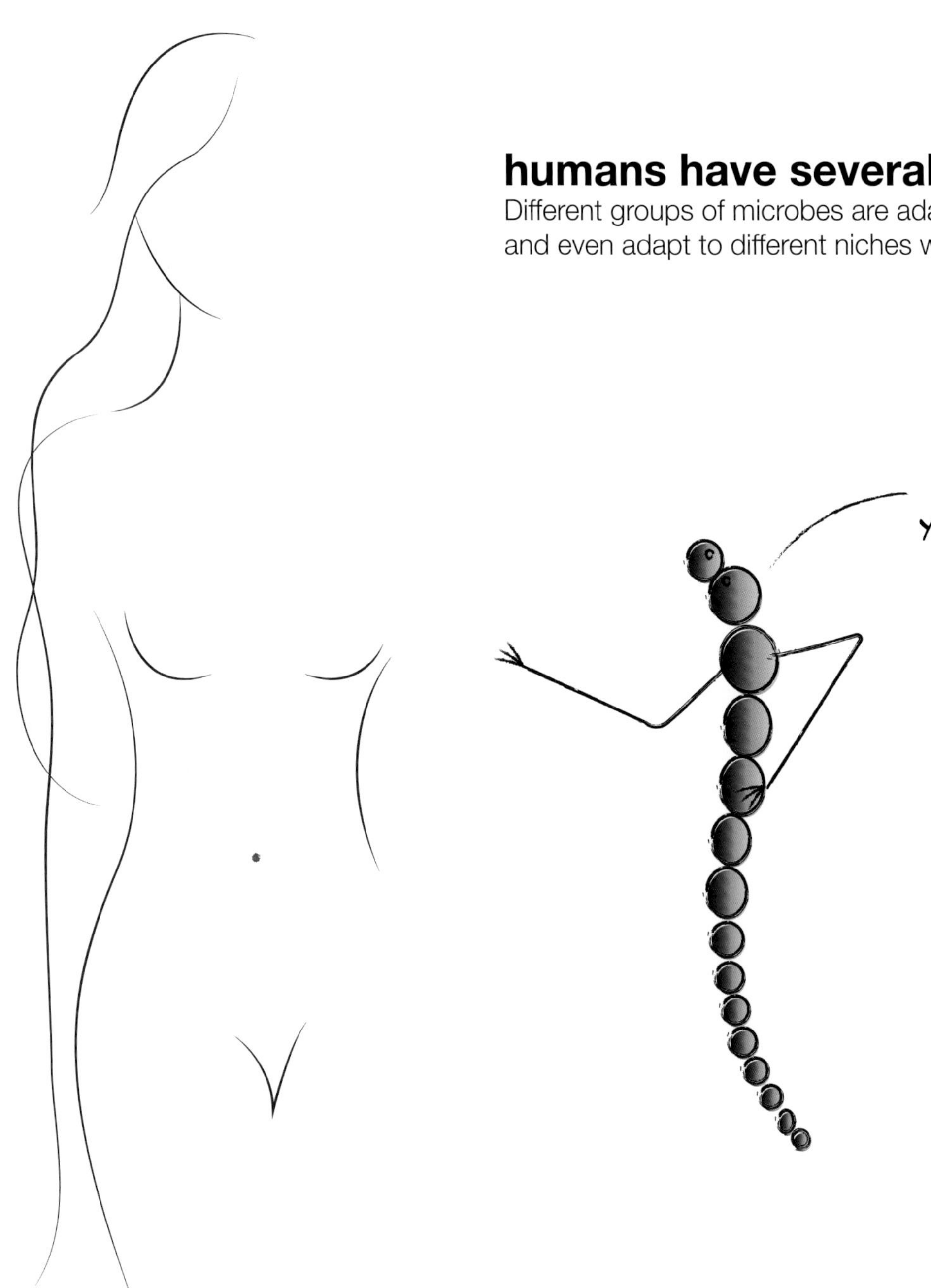

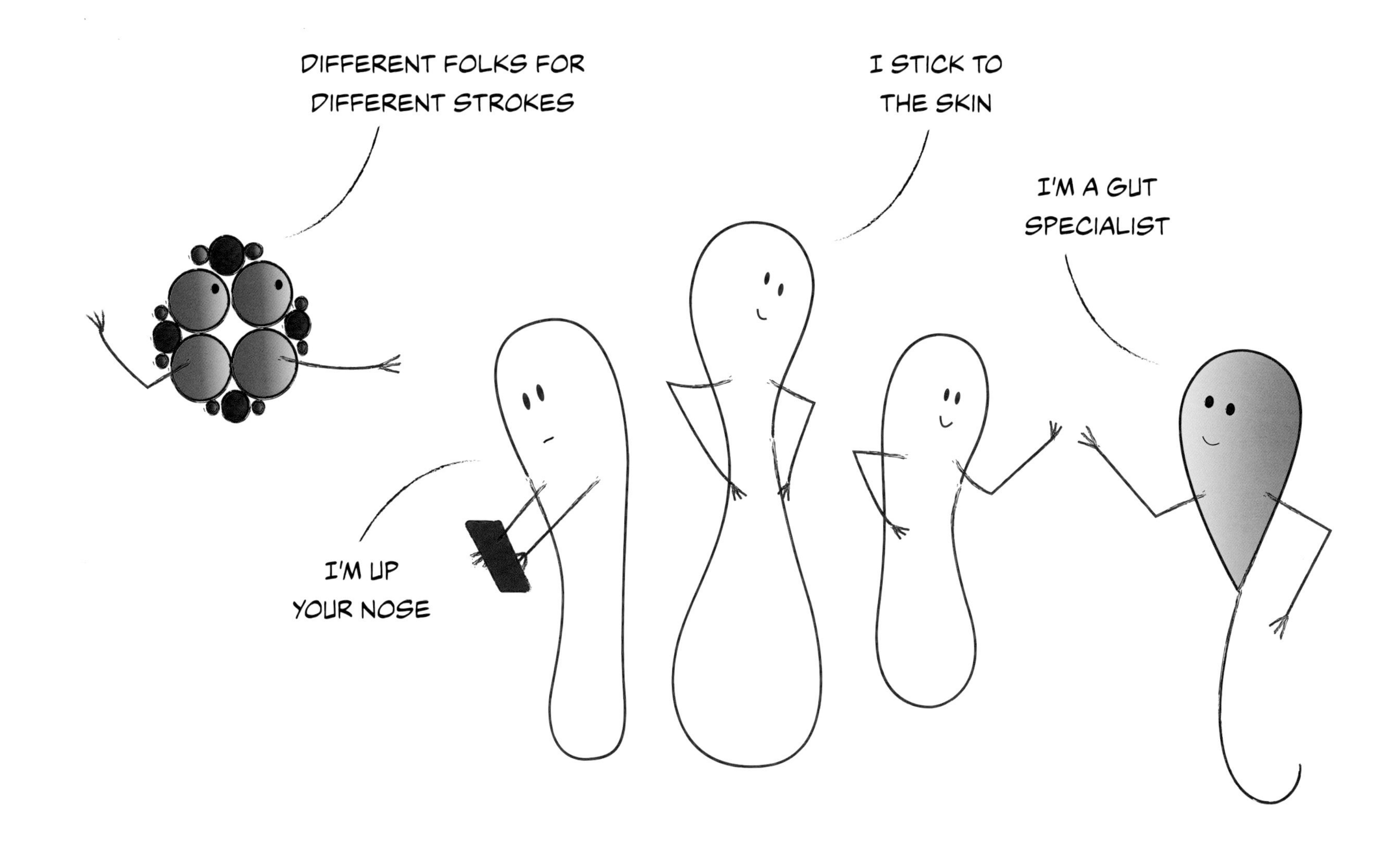
DIFFERENT FOLKS FOR
DIFFERENT STROKES
I'M UP
YOUR NOSE
I STICK TO
THE SKIN
I'M A GUT
SPECIALIST

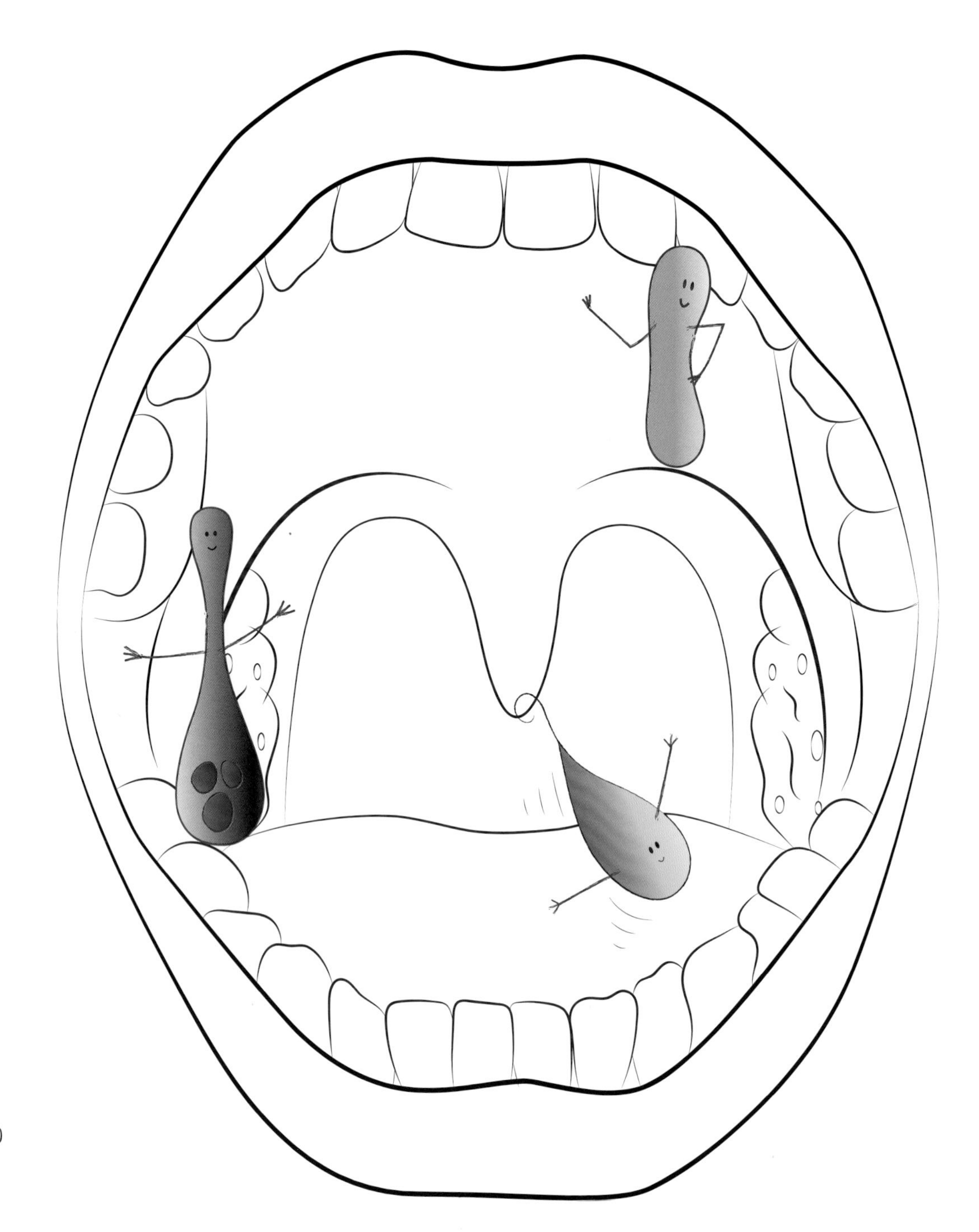

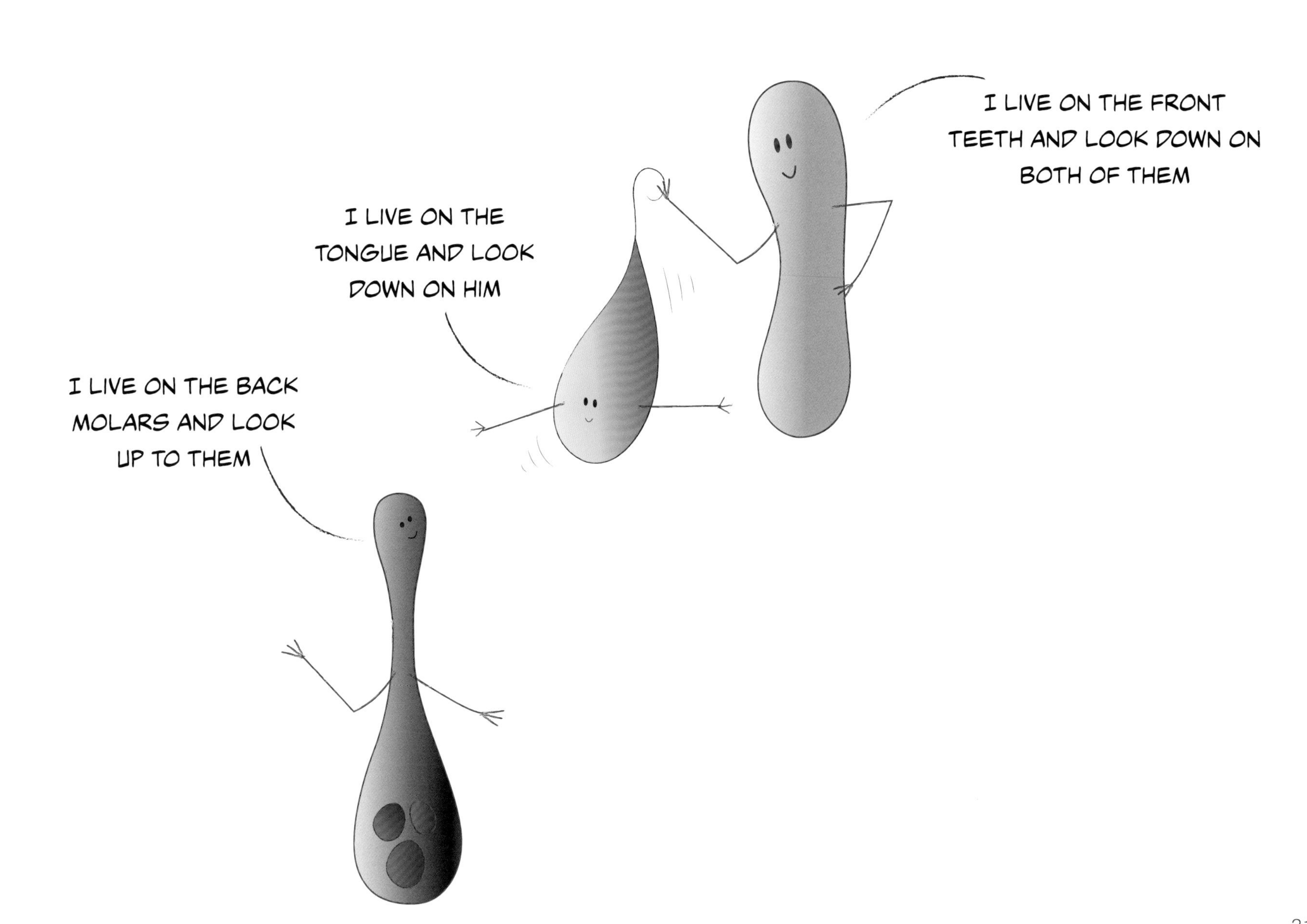
I LIVE ON THE FRONT TEETH AND LOOK DOWN ON BOTH OF THEM
I LIVE ON THE TONGUE AND LOOK DOWN ON HIM
I LIVE ON THE BACK MOLARS AND LOOK UP TO THEM

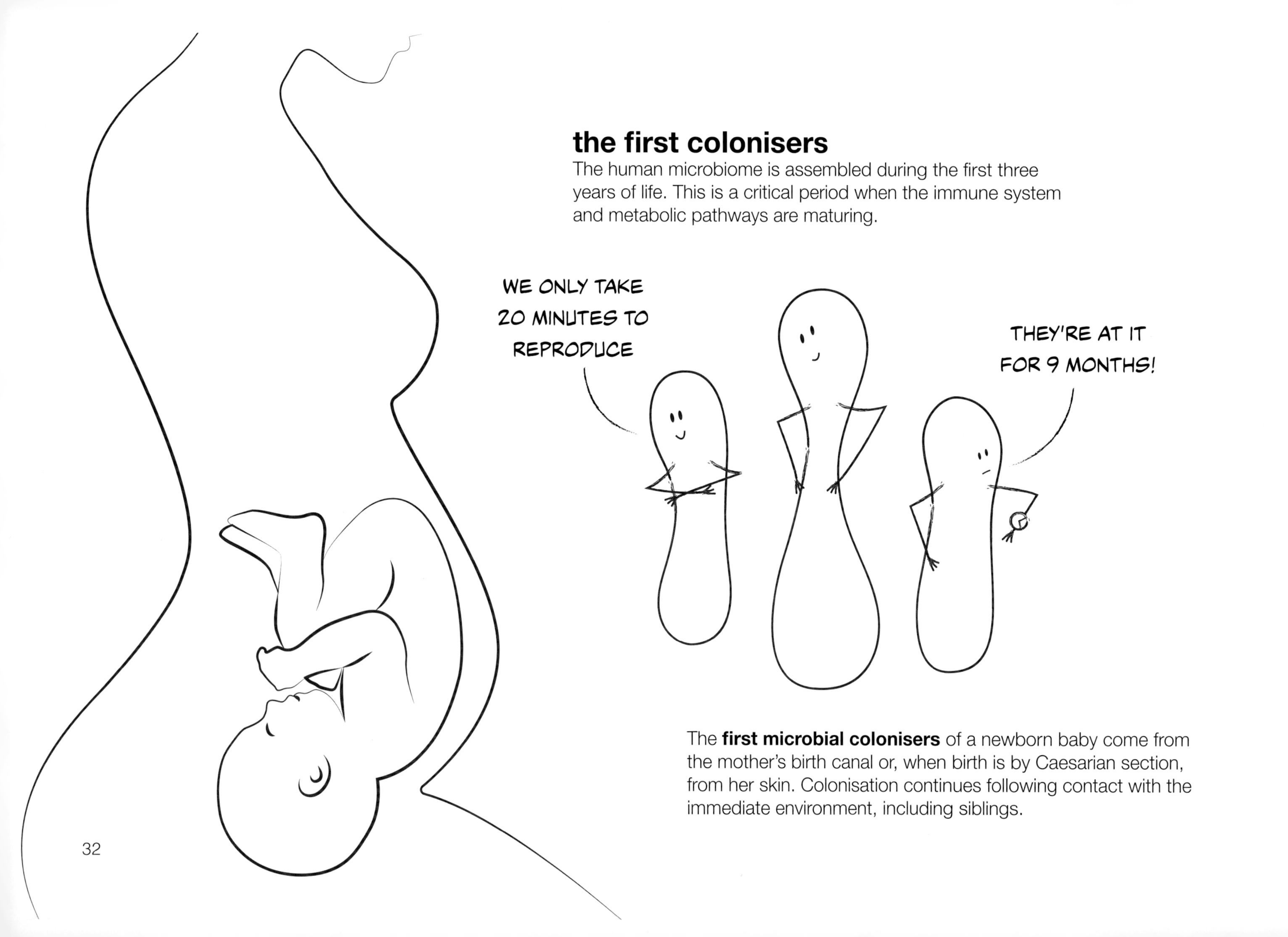

the first colonisers

The human microbiome is assembled during the first three years of life. This is a critical period when the immune system and metabolic pathways are maturing.

The **first microbial colonisers** of a newborn baby come from the mother's birth canal or, when birth is by Caesarian section, from her skin. Colonisation continues following contact with the immediate environment, including siblings.

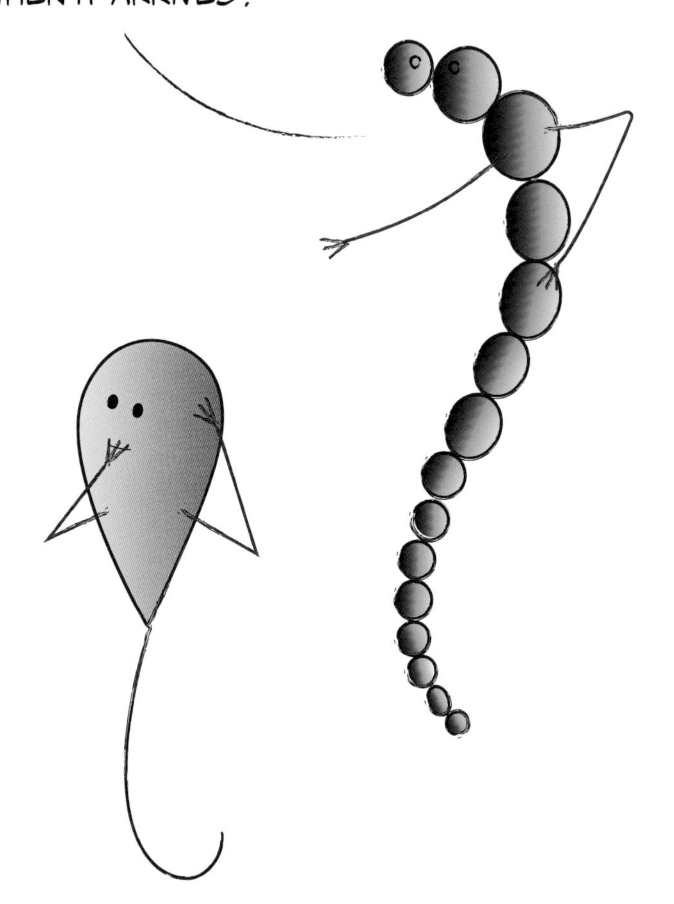
WHICH OF US WILL
HANG OUT WITH THE HUMAN
WHEN IT ARRIVES?

IT DEPENDS WHICH
WAY IT COMES OUT!

born too soon

Colonisation by microbes is a harmless asset for a fully developed newborn, but when a baby is born prematurely, its immune system, intestinal barrier and blood-brain barrier are immature; and colonisation represents a potential threat.

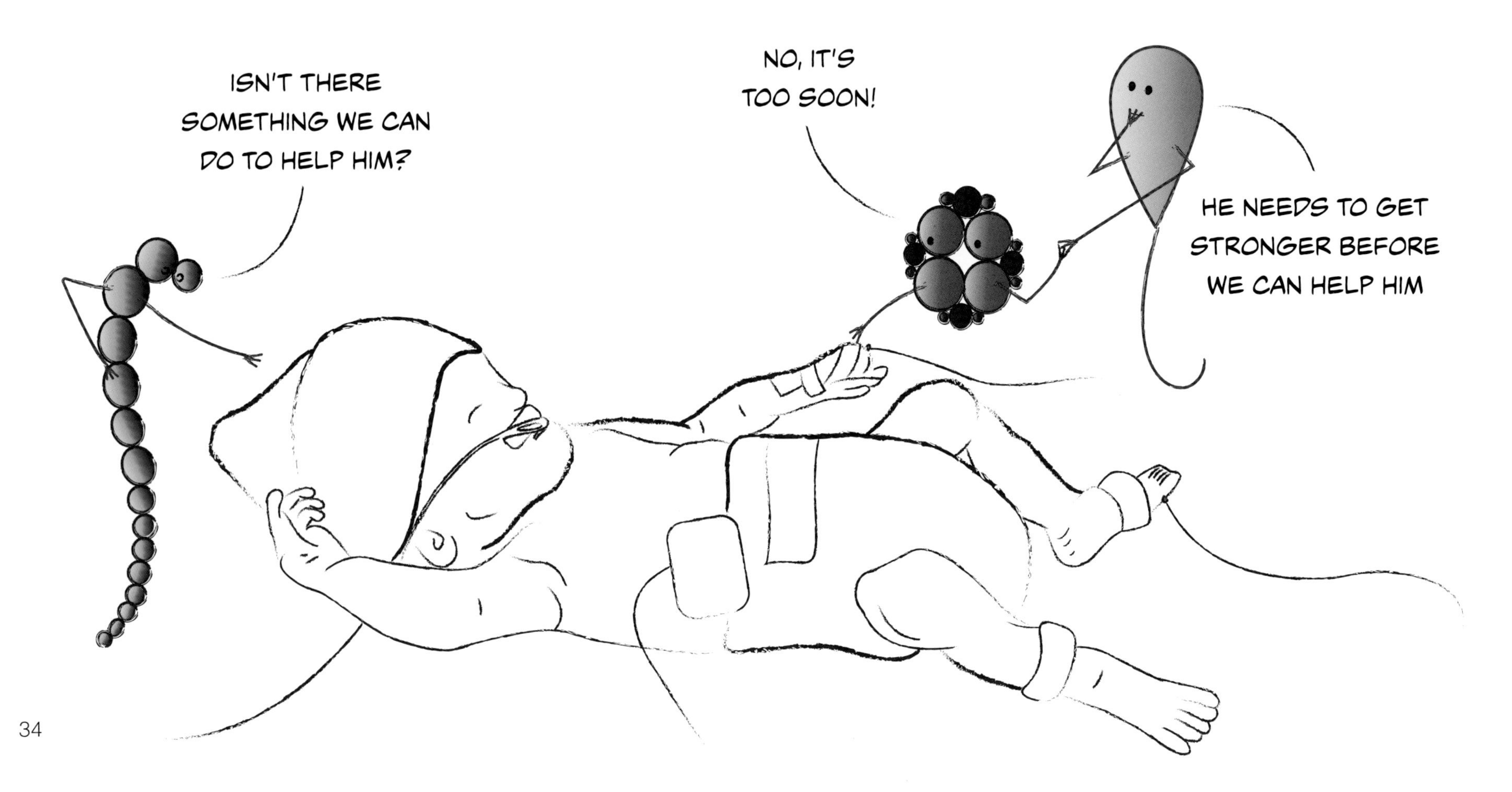

human breast milk

The assembly of the microbiome is largely completed during the first few years of life and resembles that of an adult at three to four years of age. During that period, the introduction of solid food and withdrawal of breastfeeding are important milestones and modifiers of the microbiome.

Human breast milk is one of nature's truly functional foods, meaning that the benefit it provides to an infant is more than the nutritional value of its contents. For example, human milk sugars (known as milk oligosaccharides) are food for the infant's microbes (promoting the growth of beneficial *Bifidobacteria*) while simultaneously protecting against infections and enhancing the infant's immune system.

Although the benefits of breast feeding are significant, formula-fed babies are not necessarily at any long-term disadvantage. The value of breast feeding is particularly important for premature babies.

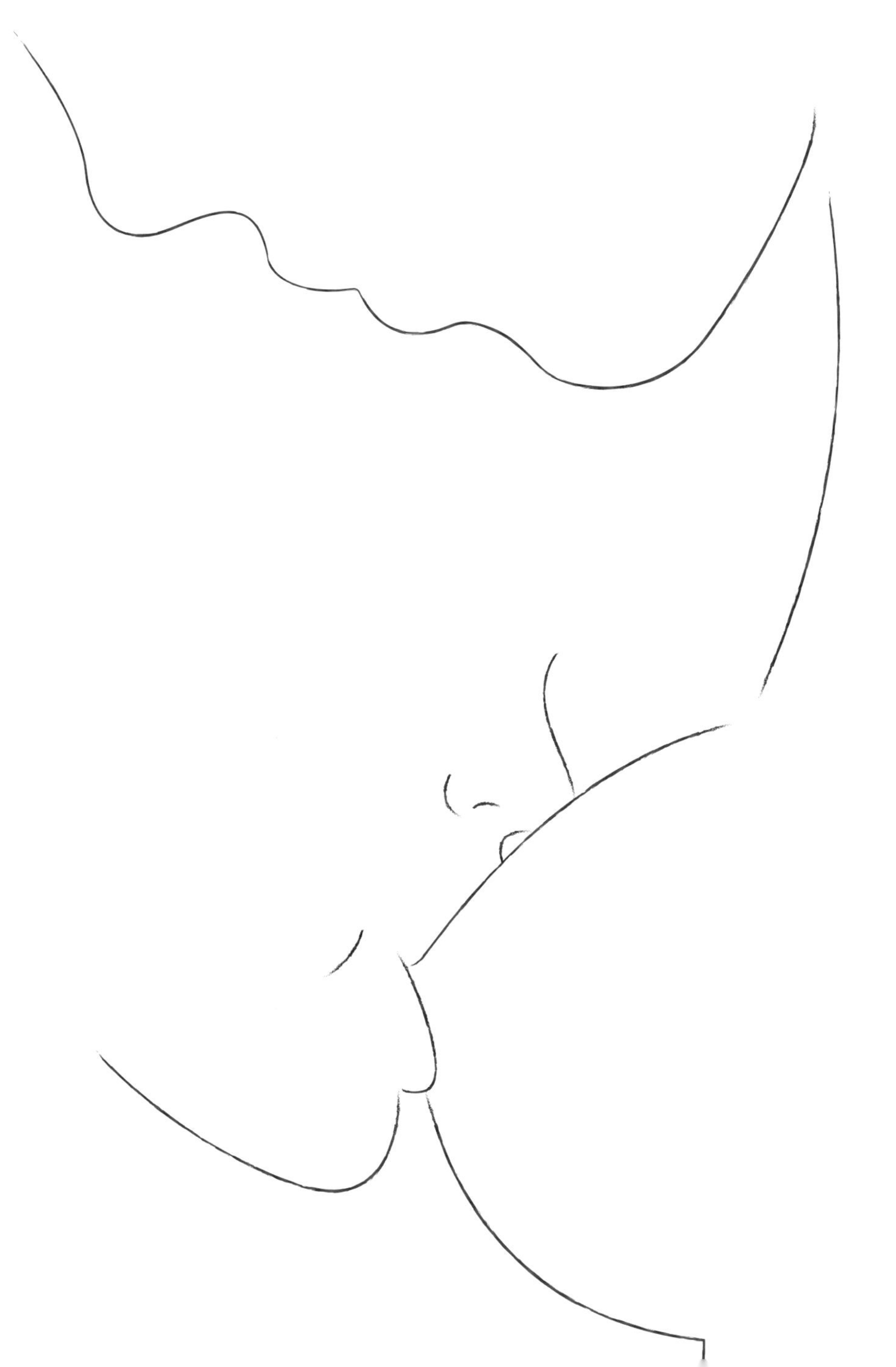

HEY BABY . . . THE GENES YOU GOT FROM YOUR PARENTS HAVE INSUFFICIENT INFORMATION FOR YOU TO DEVELOP FULLY

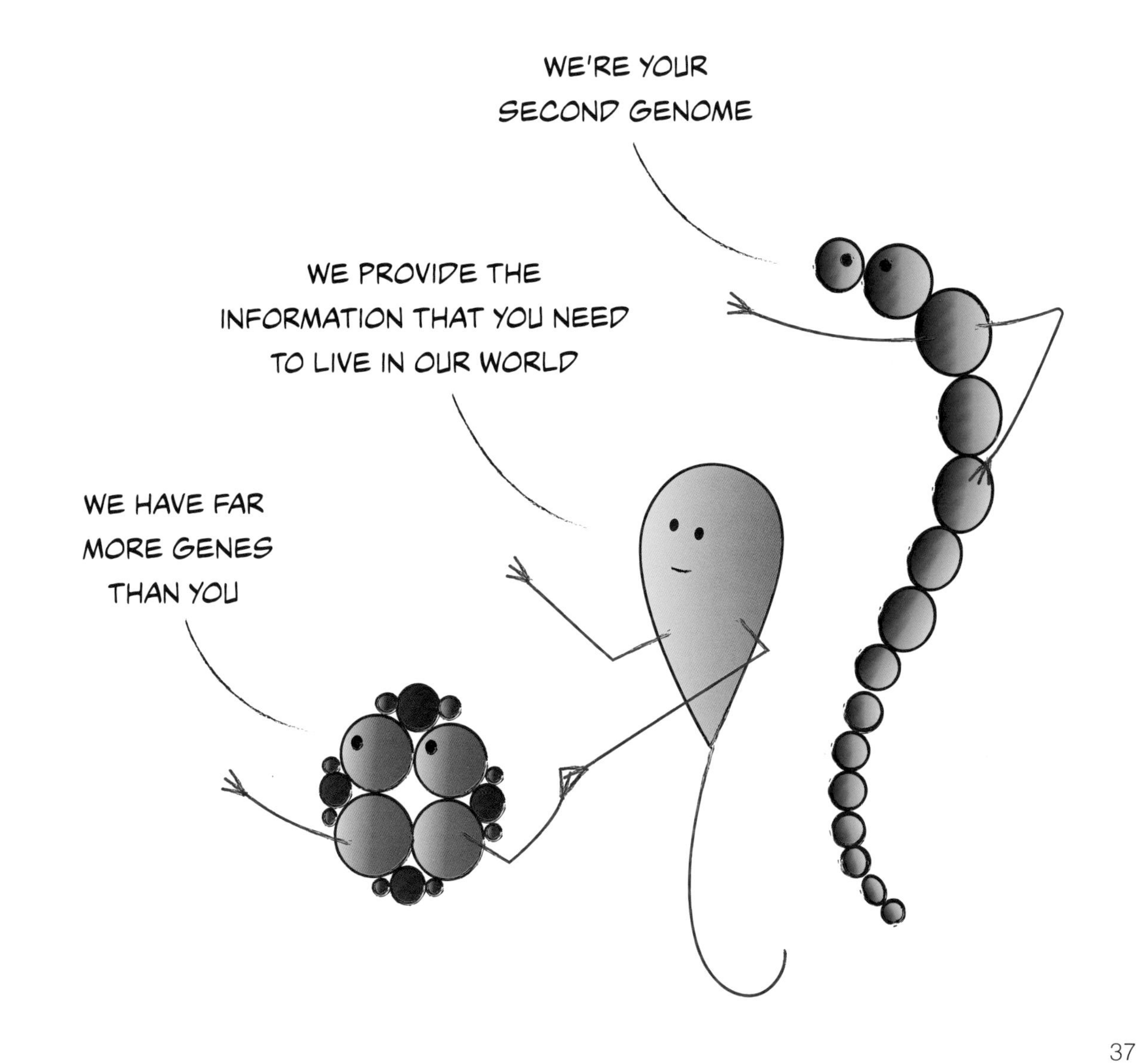
WE'RE YOUR SECOND GENOME
WE PROVIDE THE INFORMATION THAT YOU NEED TO LIVE IN OUR WORLD
WE HAVE FAR MORE GENES THAN YOU

the social microbiome

Although humans get their first microbes at birth by mother-to-baby transmission, the next wave of microbes is acquired by transmission from other people, from the food we eat, the air we breathe, animals and environmental surfaces.

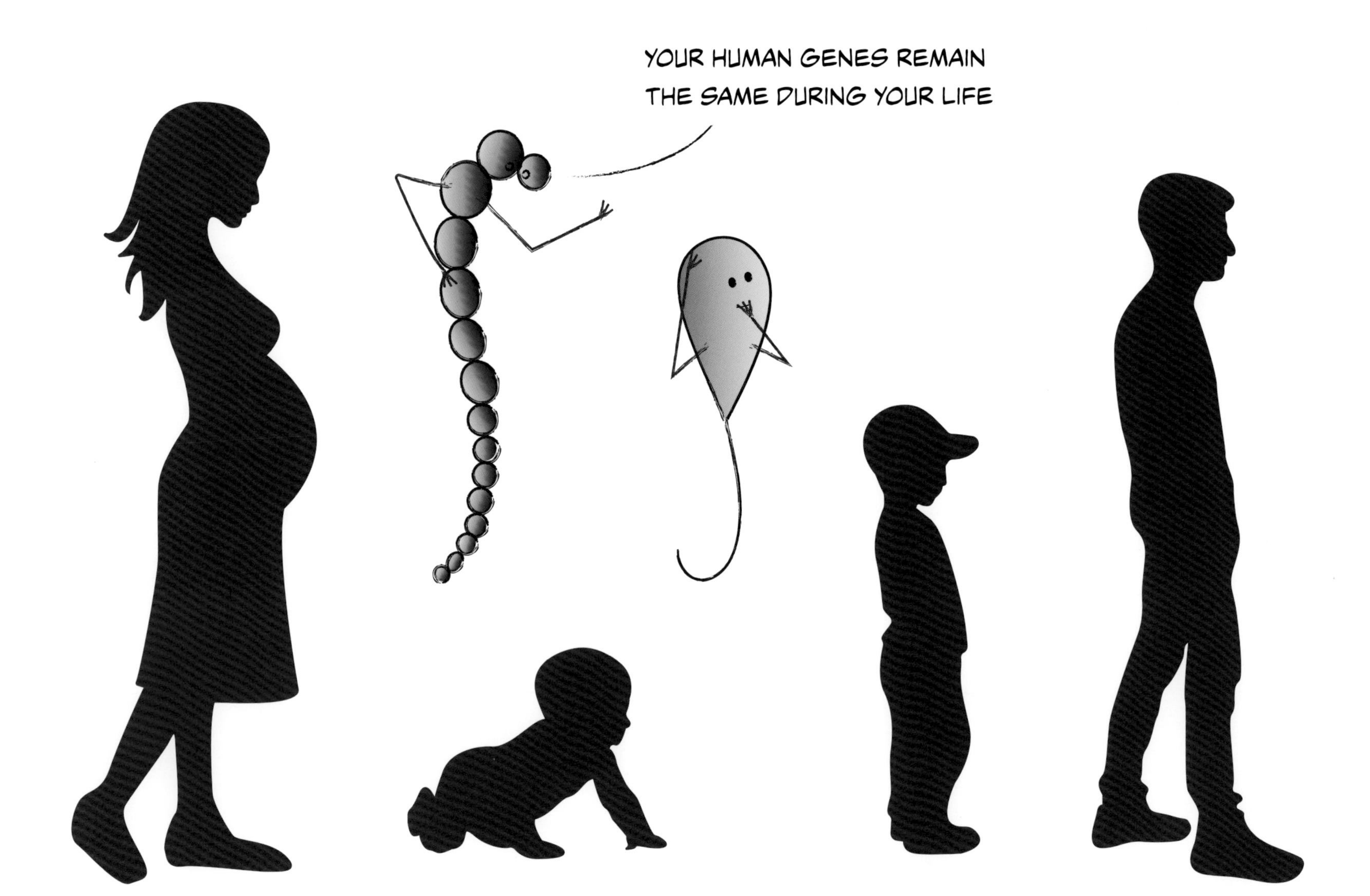
YOUR HUMAN GENES REMAIN THE SAME DURING YOUR LIFE

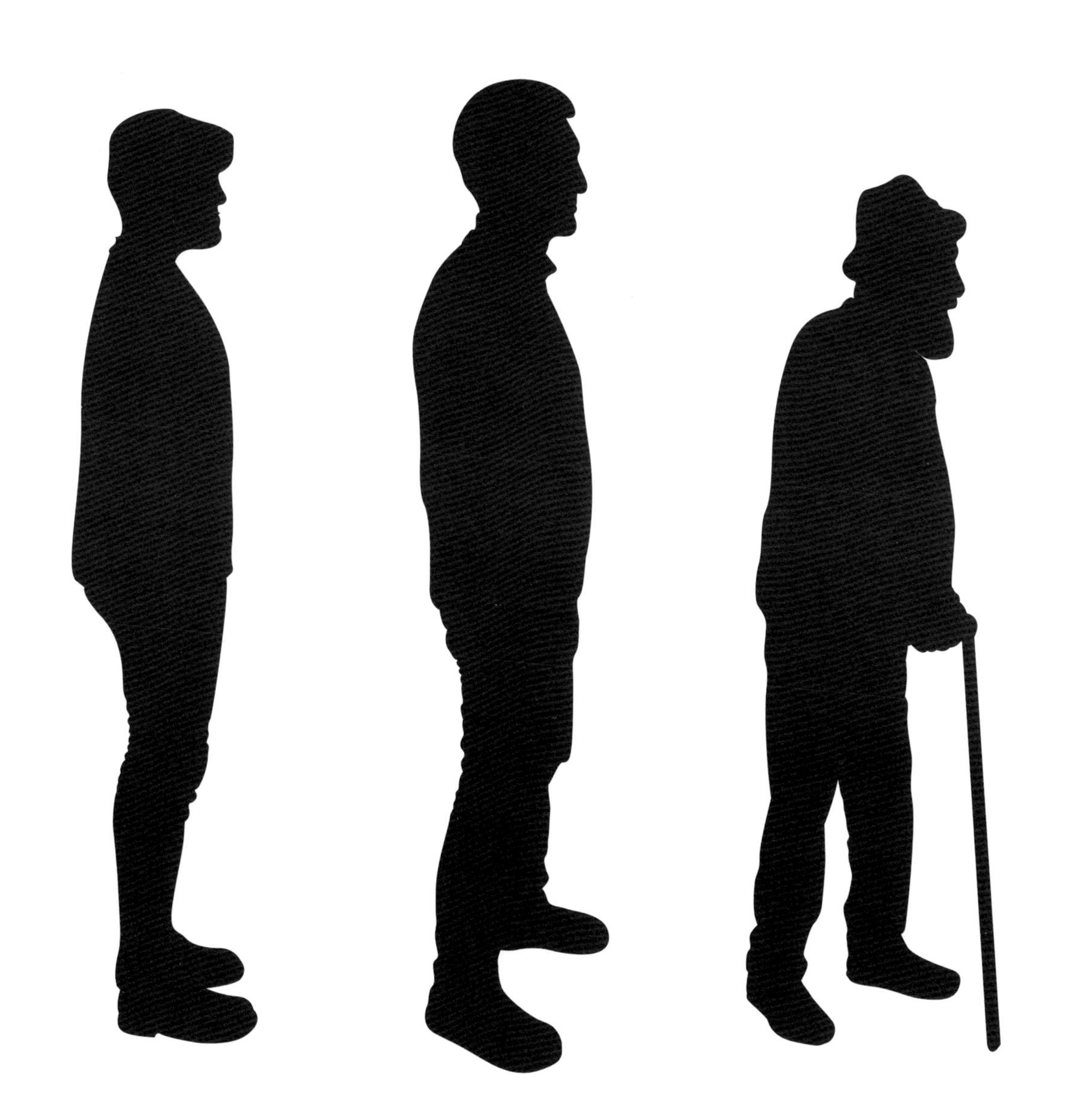

BUT YOUR
MICROBIOME ADAPTS TO
CHANGING TIMES

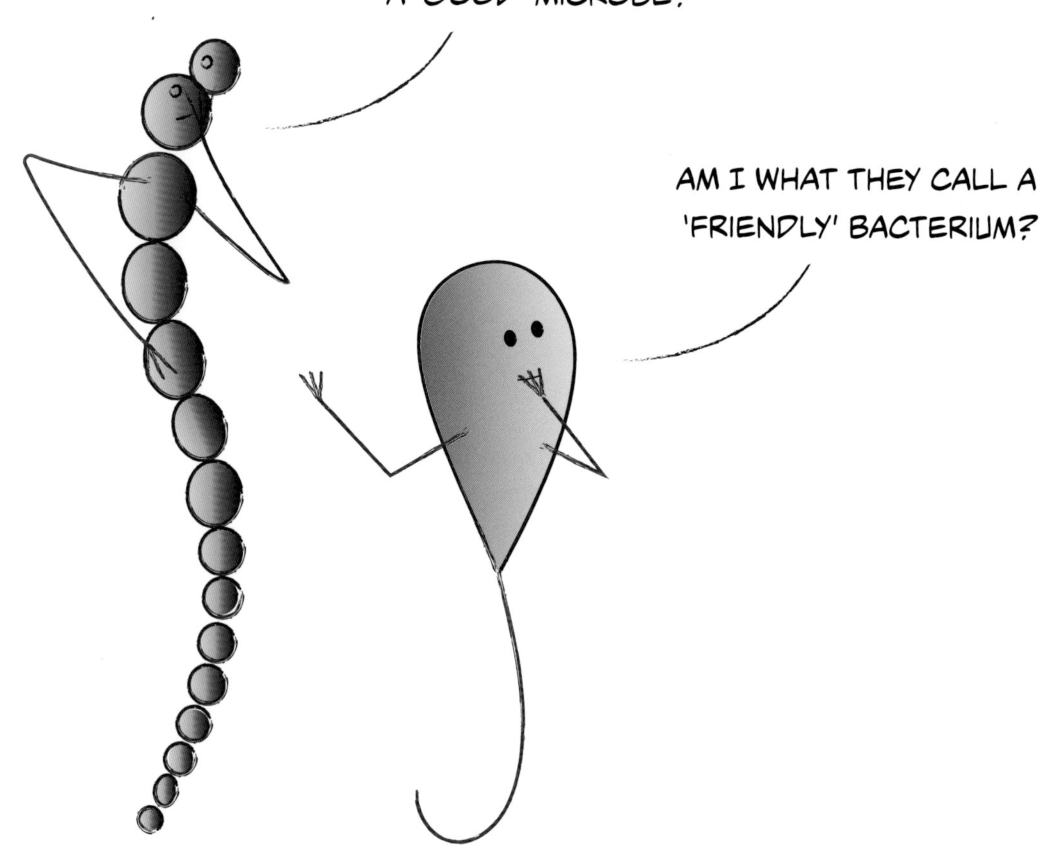
AM I WHAT HUMANS CALL A 'GOOD' MICROBE?
AM I WHAT THEY CALL A 'FRIENDLY' BACTERIUM?

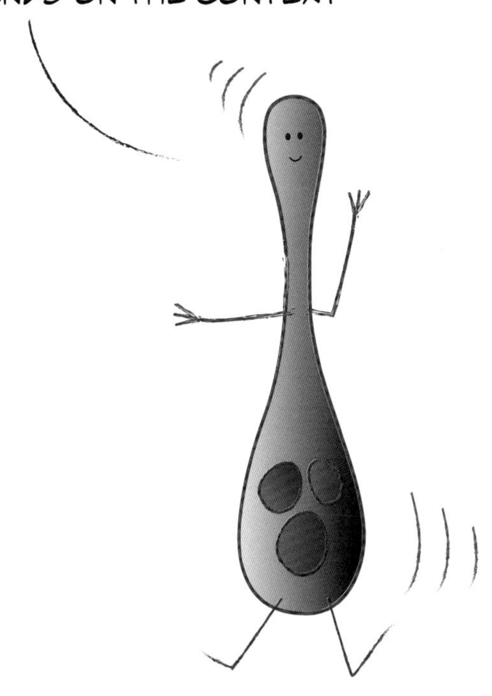
HUMAN VALUES DON'T APPLY TO MICROBES. 'GOOD' OR 'BAD' DEPENDS ON THE CONTEXT

THERE'S NO SUCH THING AS 'GOOD' OR 'BAD'OR 'FRIENDLY' MICROBES. WE JUST EXIST AND CONTINUALLY ADAPT TO THE CONDITIONS WE FIND OURSELVES IN

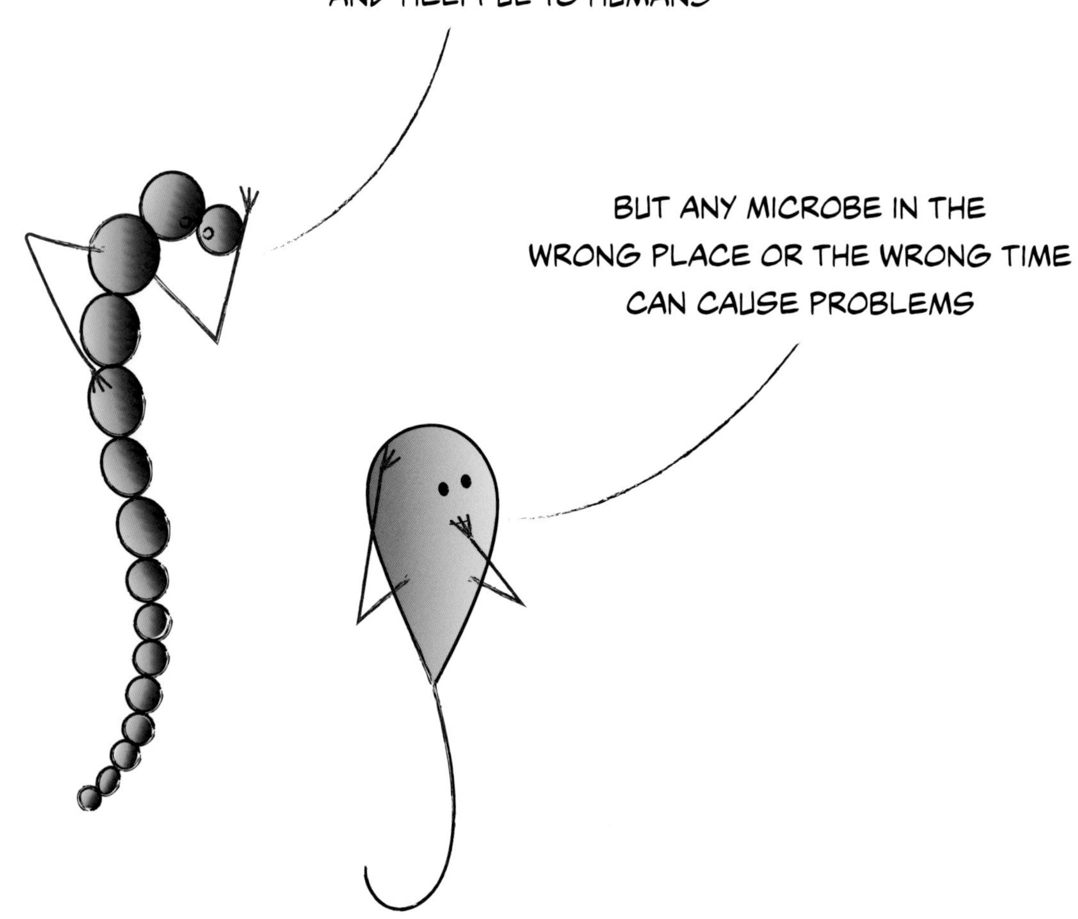
MOST MICROBES ARE HARMLESS AND HELPFUL TO HUMANS
BUT ANY MICROBE IN THE WRONG PLACE OR THE WRONG TIME CAN CAUSE PROBLEMS

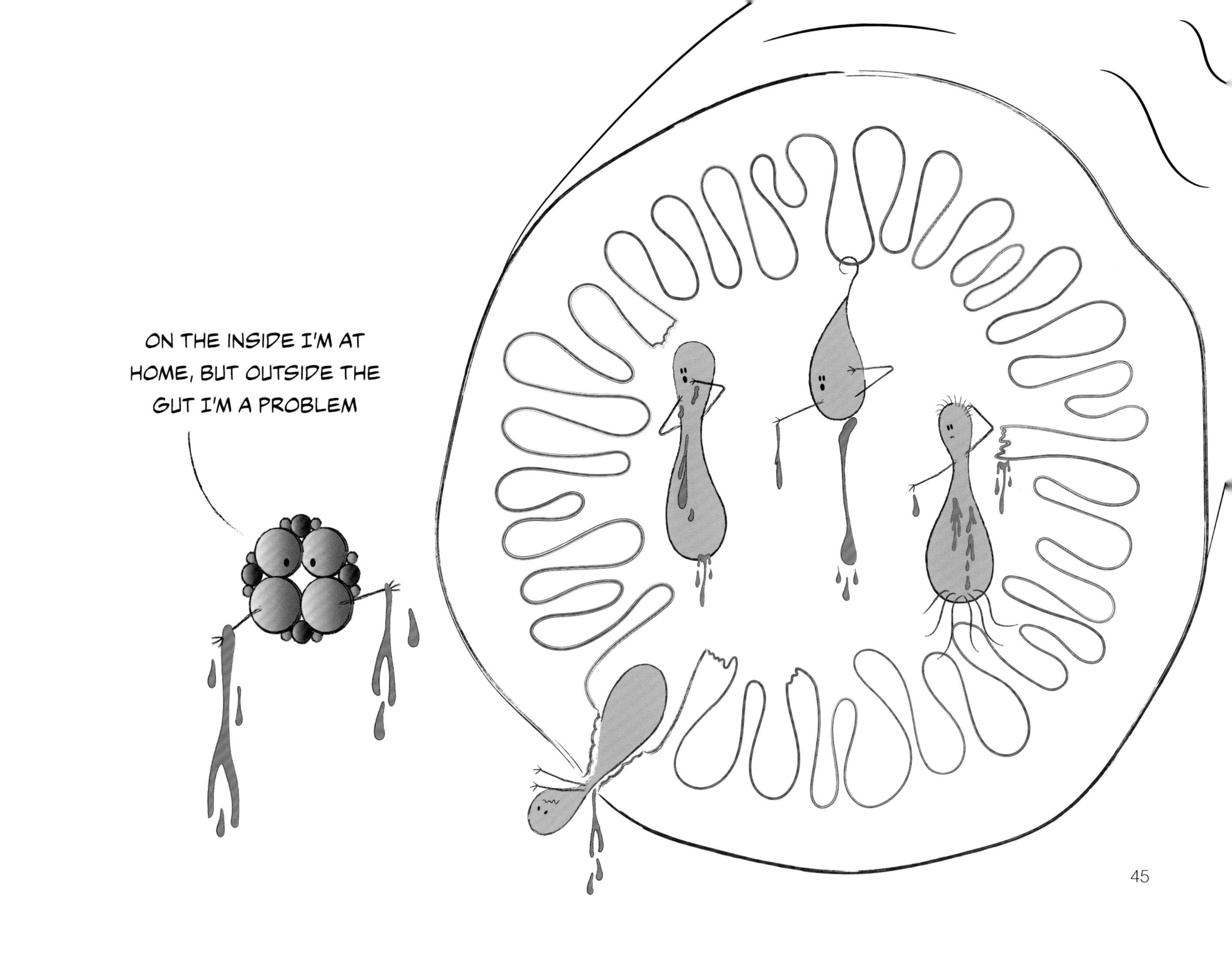
ON THE INSIDE I'M AT HOME, BUT OUTSIDE THE GUT I'M A PROBLEM

By way of example, ***Helicobacter pylori*** may cause ulcers and stomach cancer in some people but also protects against cancer of the lower oesophagus and other chronic diseases. Historically, it may have given humans a survival advantage, but its association with humans has been in decline for decades in industrialised countries.

James Joyce (1882-1941) complained throughout his adult life of 'duodismal' symptoms and died aged fifty-eight from complications of a duodenal ulcer probably caused by *Helicobacter pylori*.

strains matter

Strains of the same bacterial species look alike but behave differently. For example, one of the most studied bacterial species is *Escherichia coli (E. coli)*. Some strains of *E. coli* promote food digestion whereas others may cause food poisoning and inflammatory disease, and still others have been used as a health-promoting probiotic. Many other types of microbe in the gut have similar varying functions depending on the strain.

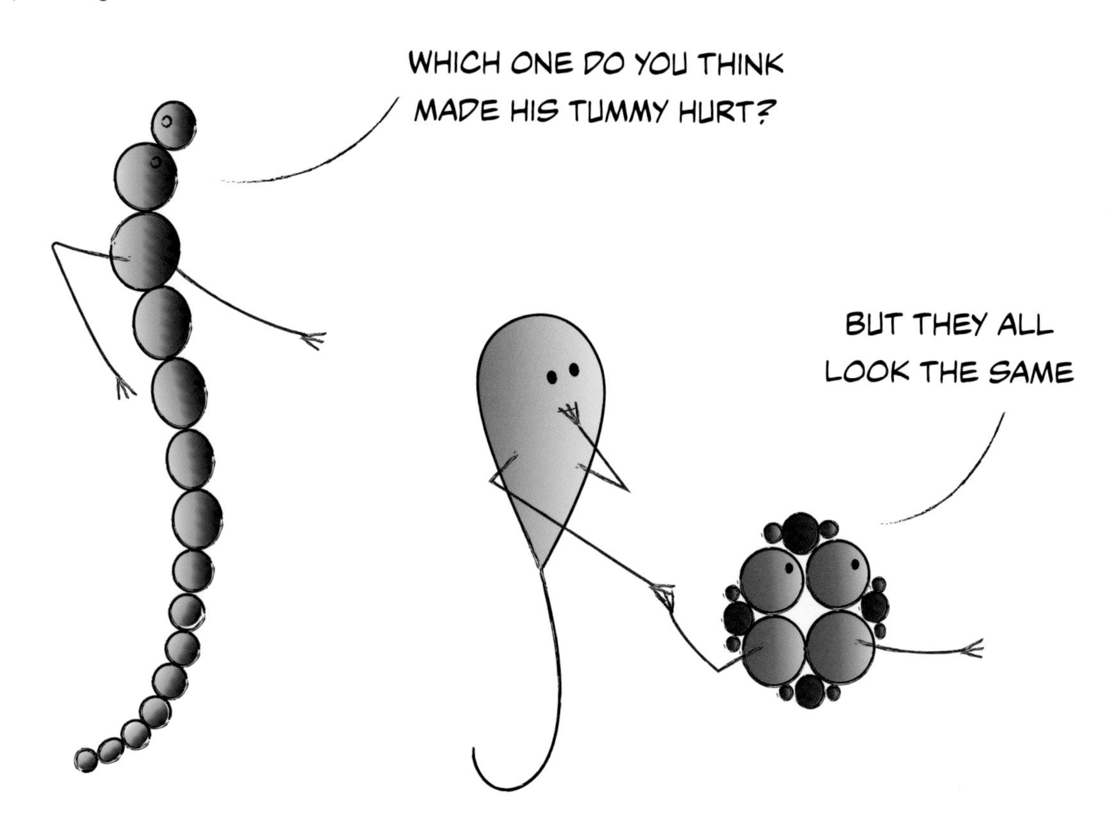

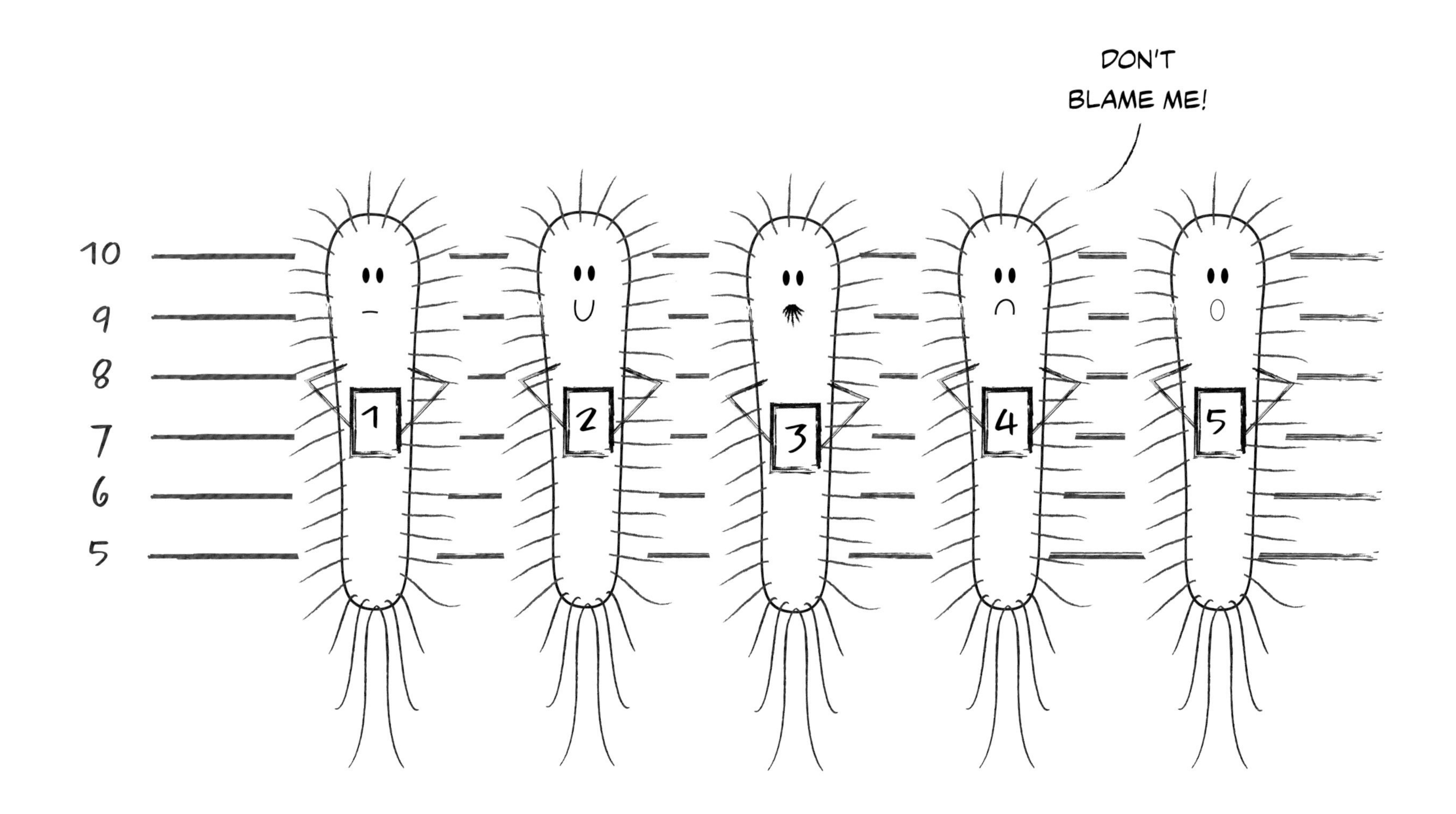
DON'T BLAME ME!
10
9
8
7
6
5
1
2
3
4
5

what's sex got to do with it?

Single-cell microorganisms reproduce not by sex but by replicating their chromosome and dividing into daughter cells, each genetically identical to the parent cell. This asexual reproduction leads to a large population of identical cells but they have strategies for diversifying by taking up microbial genes from the environment and by transferring genes, including antibiotic-resistance genes, to other bacteria by direct contact (conjugation). Viruses (bacteriophages) may also transfer genes from one bacterium to another.

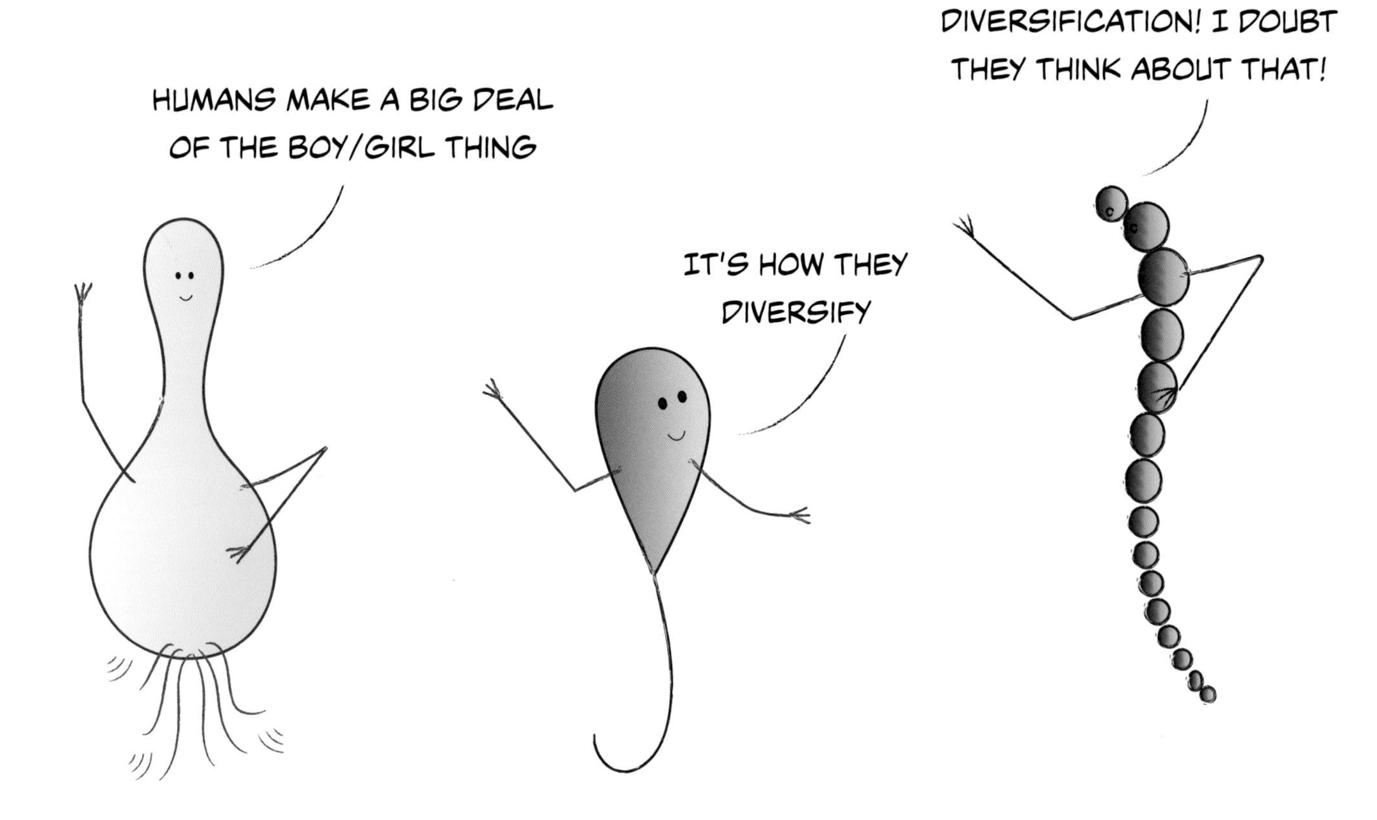

The sex of the host influences the composition and function of the microbiota due in part to how microbes interact with female and male sex hormones.

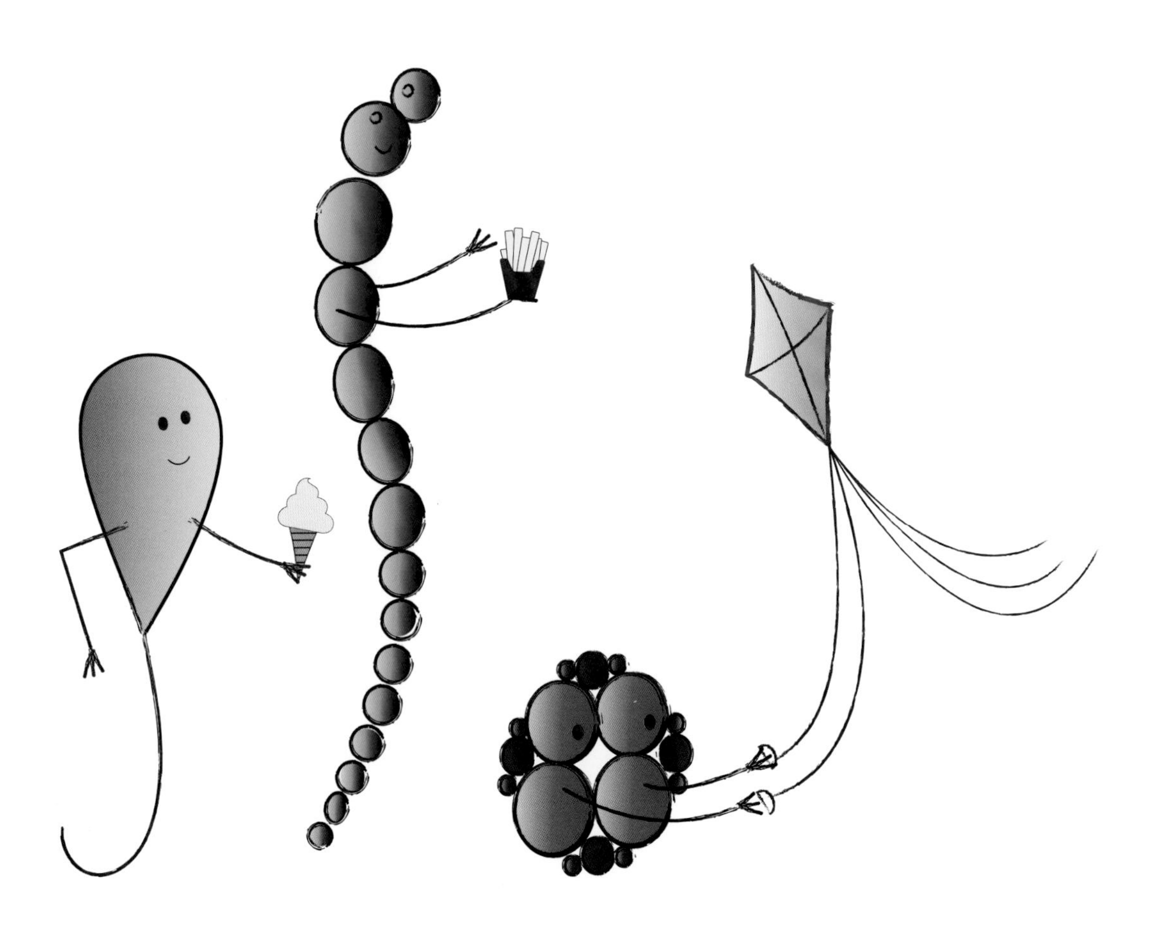

3 living together

When evolution matched humans with their microbes, it created a wonderful partnership. Microbes want an environment where they can survive. In exchange, they provide protection and assistance with digestion, nutrition and metabolism.

Like all good partnerships, there is continual dialogue, with chemical signals going from microbe to host and back again to ensure that the two of them work well together. This happens silently – we are not aware of it happening – but every day gut microbes send a reminder that they are still there.

Read on to find out what this signal is!

mind your microbes

Almost every aspect of human lifestyle affects gut microbes. Food, sleep, drugs, physical and social activity all influence the composition and function of gut microbes. In turn, the microbiota reflect human health and well-being.

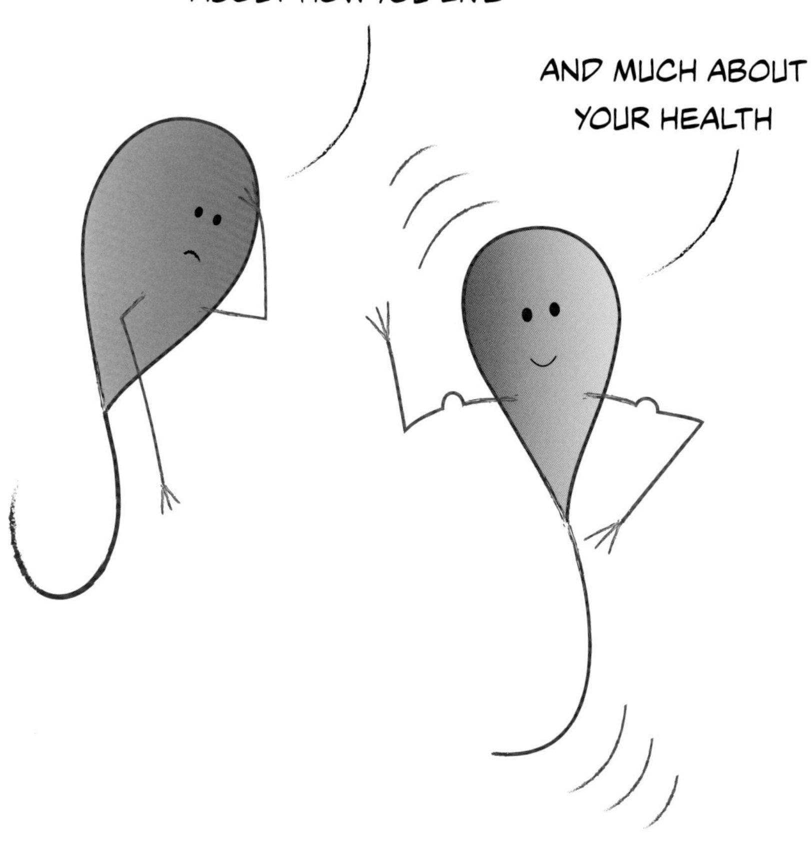

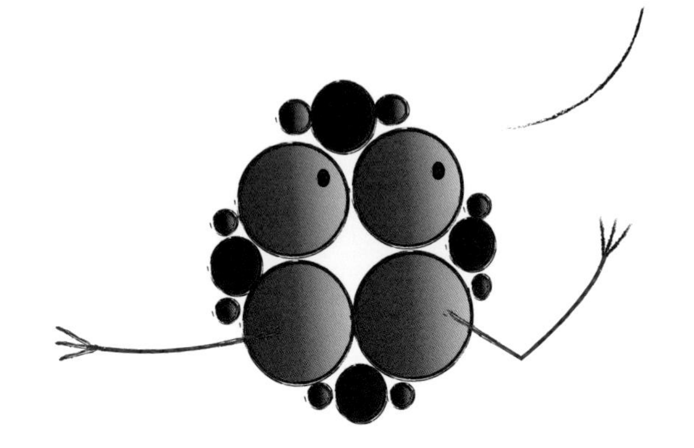

WHY AM I SO FRAZZLED?

STRESS AFFECTS EVERY PART OF THE BODY - AND US TOO

WHY ARE YOU SO FULL OF ENERGY?

MY HOST SLEEPS WELL, SO I FEEL WELL!

microbes have rhythm

The abundance, composition and activity of gut microbes oscillate during the twenty-four-hour sleep and wakefulness cycle of their host. This allows them to rest while the host sleeps, and to vary with the cycle of food consumption and movement of food through the gut. This daily rhythm of gut microbes is disturbed in shift workers and in people with jet lag, in whom it may upset bowel function.

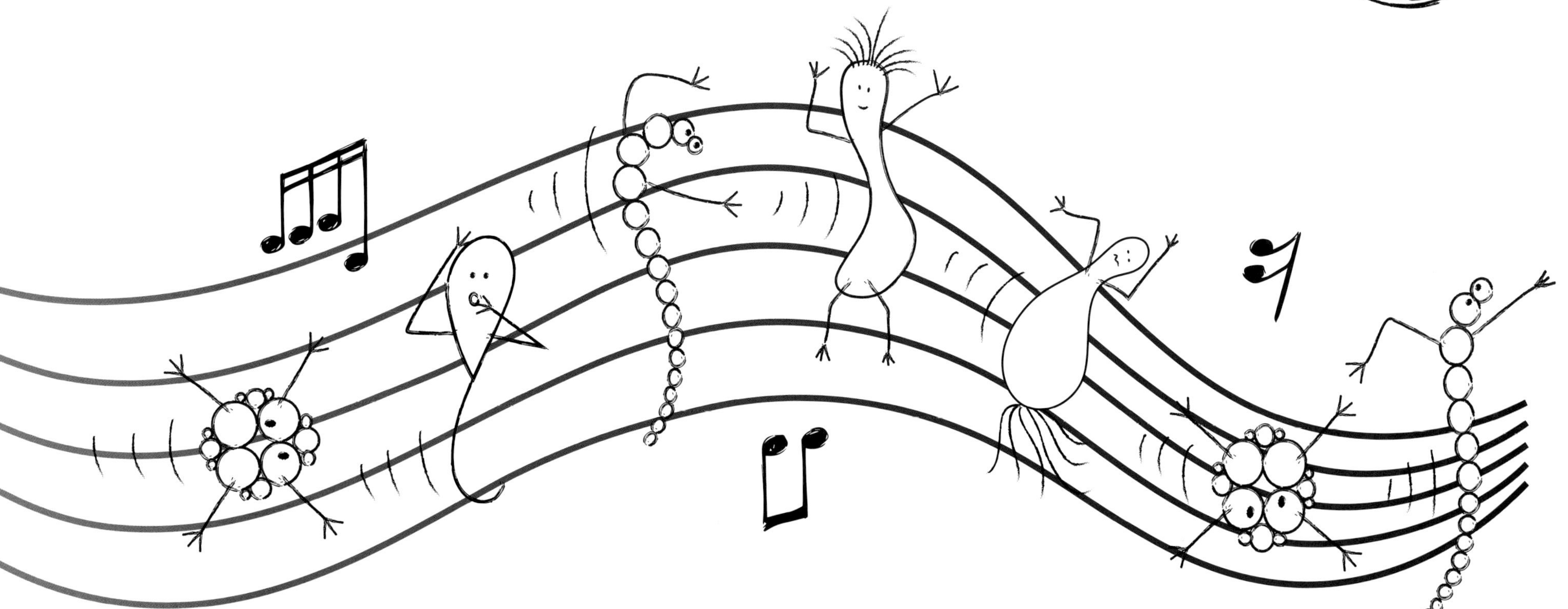

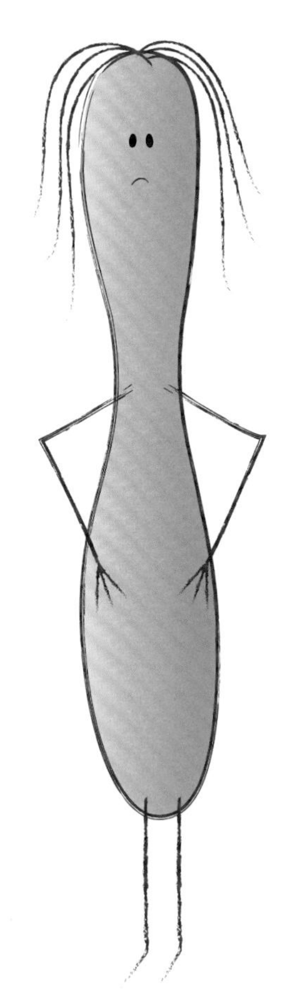
YOUR HAIR LOOKS SO LOVELY!

IT'S ALL DOWN TO MY HOST. I'M WELL FED!

WE DON'T NEED MUCH -
JUST THE RIGHT FOOD, SLEEP,
MOVEMENT AND REST!

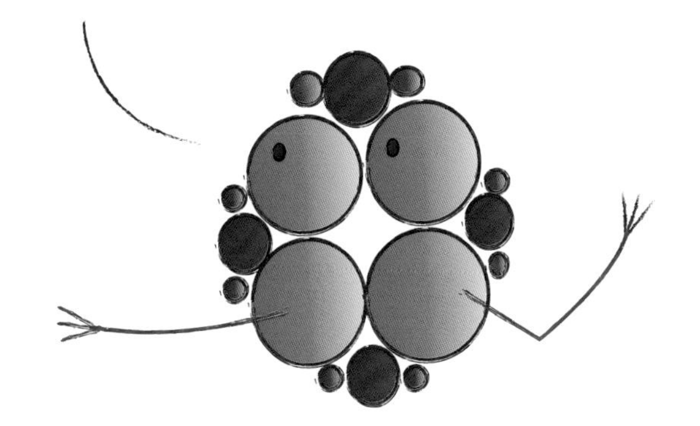

WE NEED FIBRE
OF VARIOUS TYPES
WHEN HUMANS EAT, THEY'RE
NOT JUST FEEDING THEMSELVES . . .
THEY ARE ALSO FEEDING US

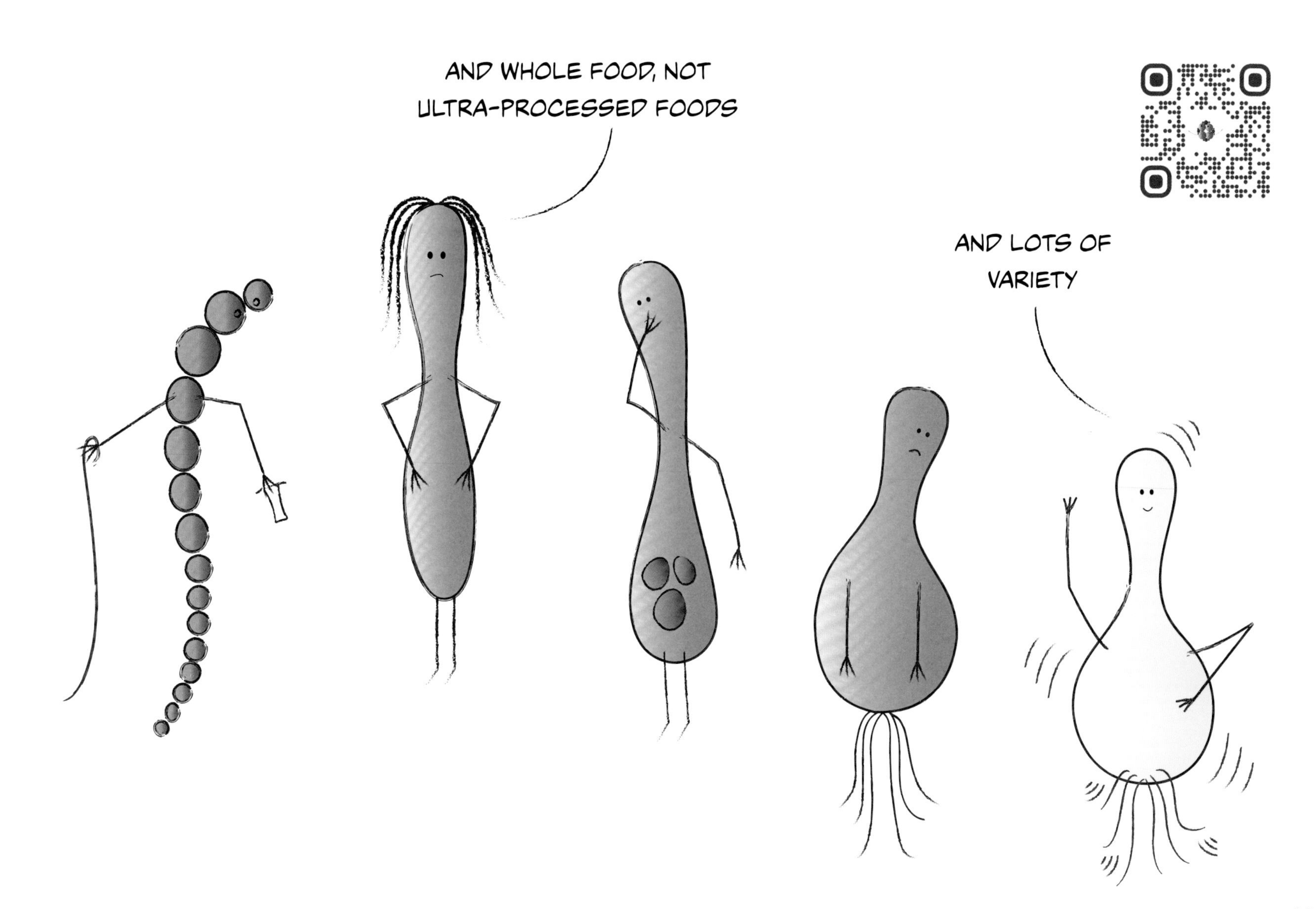
AND WHOLE FOOD, NOT
ULTRA-PROCESSED FOODS
AND LOTS OF
VARIETY

YOU CAN'T JUST
KOMBUCHA-CHA YOUR WAY TO
A HEALTHY LIFE
WHAT ARE WE HAVING
FOR TEA TONIGHT?

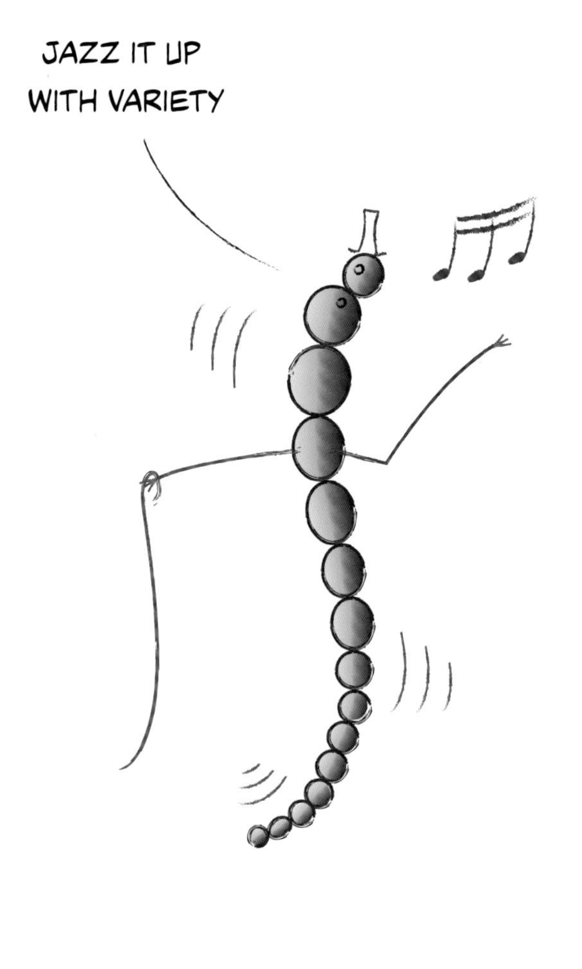
JAZZ IT UP
WITH VARIETY

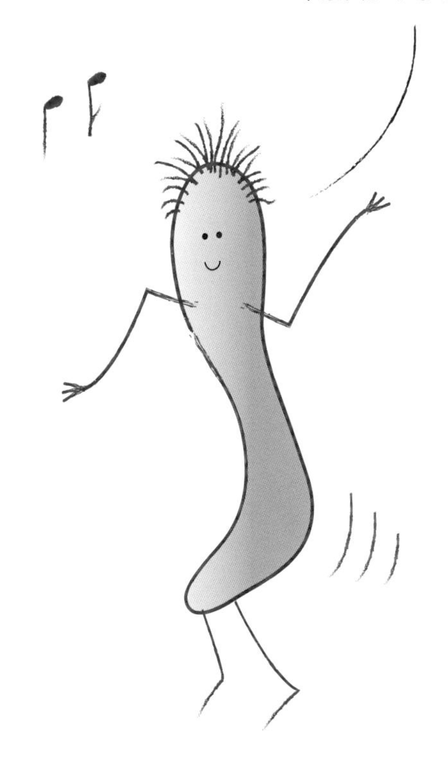
WE JIVE ON
'REAL' FOOD!

SHAKE IT UP
WITH COLOUR

KEEP
MOVIN'

eat more fibre

Just as people and microbes are not all created the same, similarly there are many different types of dietary fibre. Microbes vary in their preferences for different fibre types. A prudent diet for gut microbes is one that contains a variety of fibres.

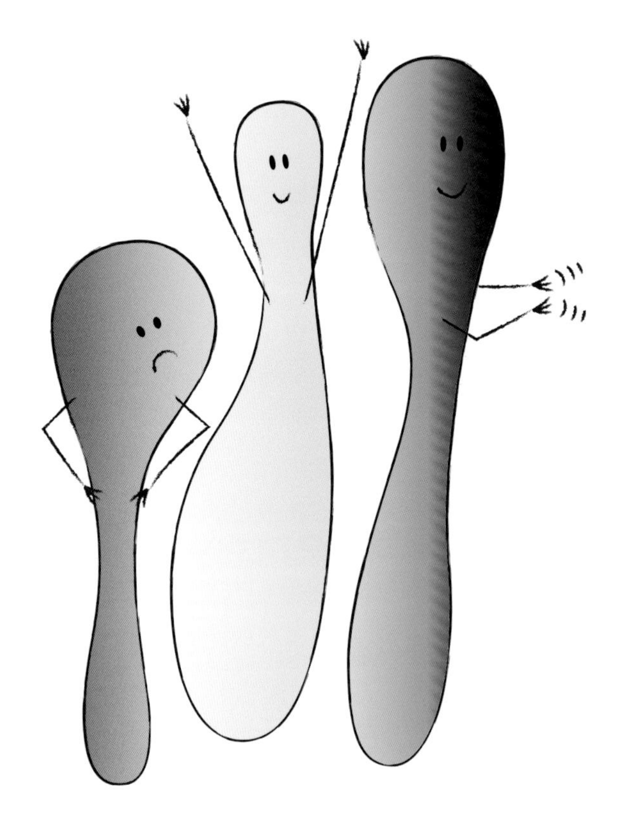

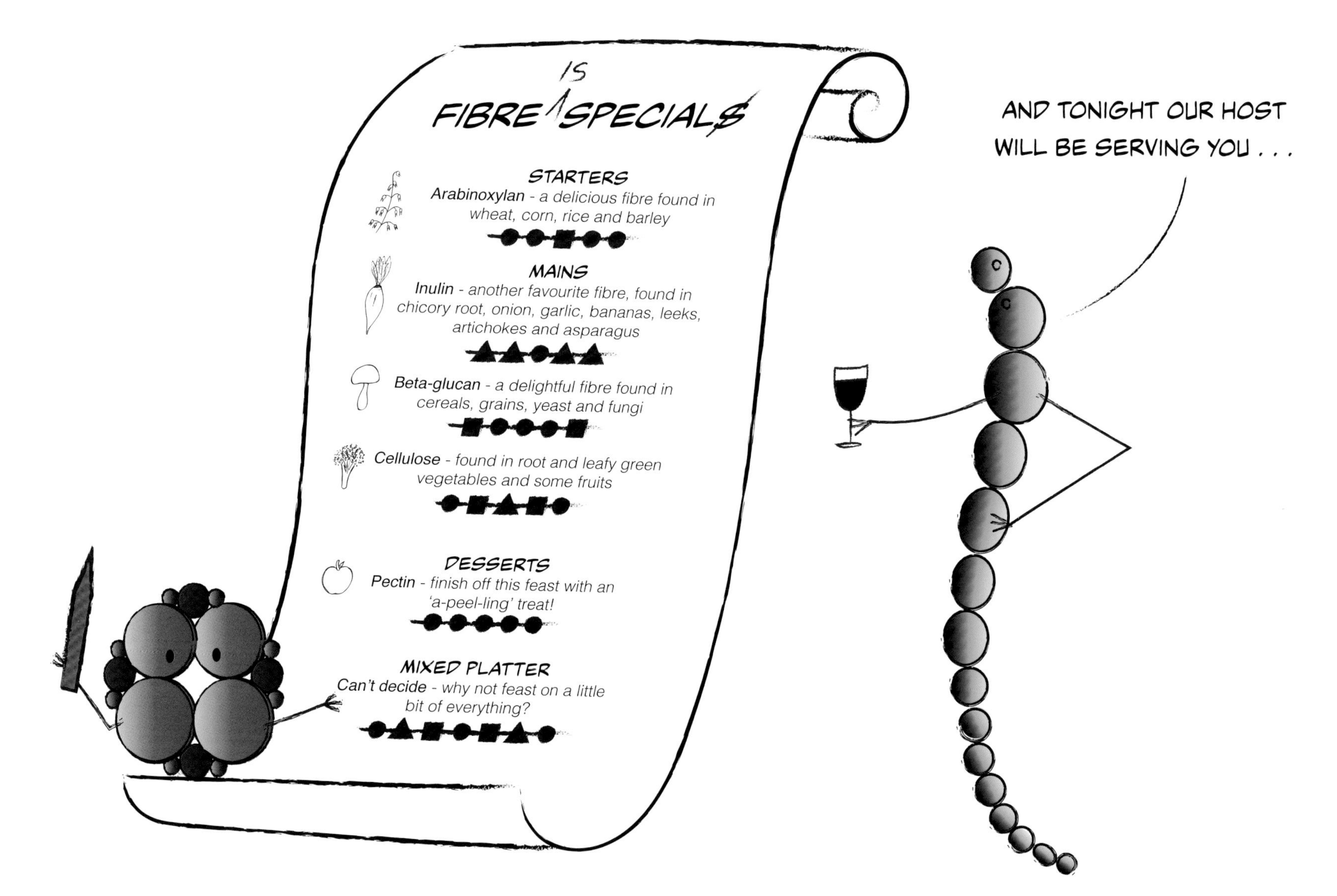
FIBRE IS SPECIALS
STARTERS
Arabinoxylan - a delicious fibre found in wheat, corn, rice and barley
MAINS
Inulin - another favourite fibre, found in chicory root, onion, garlic, bananas, leeks, artichokes and asparagus
Beta-glucan - a delightful fibre found in cereals, grains, yeast and fungi
Cellulose - found in root and leafy green vegetables and some fruits
DESSERTS
Pectin - finish off this feast with an 'a-peel-ling' treat!
MIXED PLATTER
Can't decide - why not feast on a little bit of everything?
AND TONIGHT OUR HOST WILL BE SERVING YOU . . .

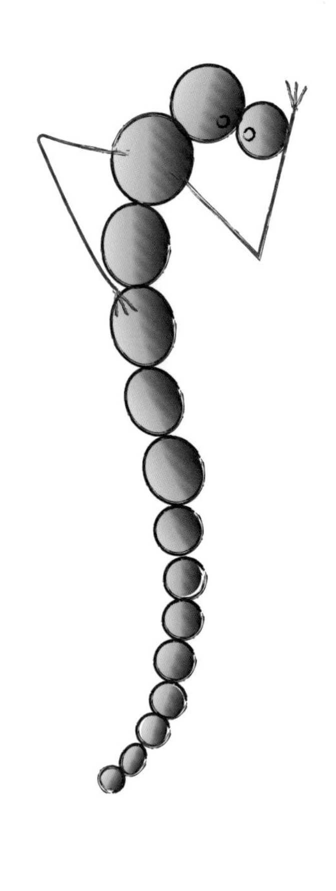

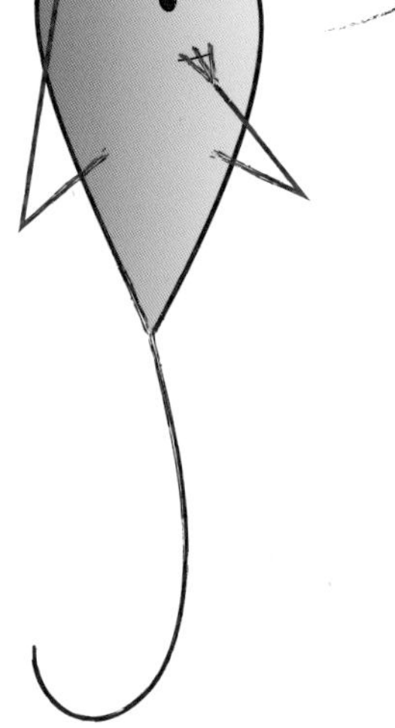

Low-fibre diets mean that microbes must adapt and find another source of food in their microenvironment, such as mucin. Mucins are sugar-coated proteins that form the gel (mucus) that covers the surface of mucous membranes such as the inner lining of the gut.

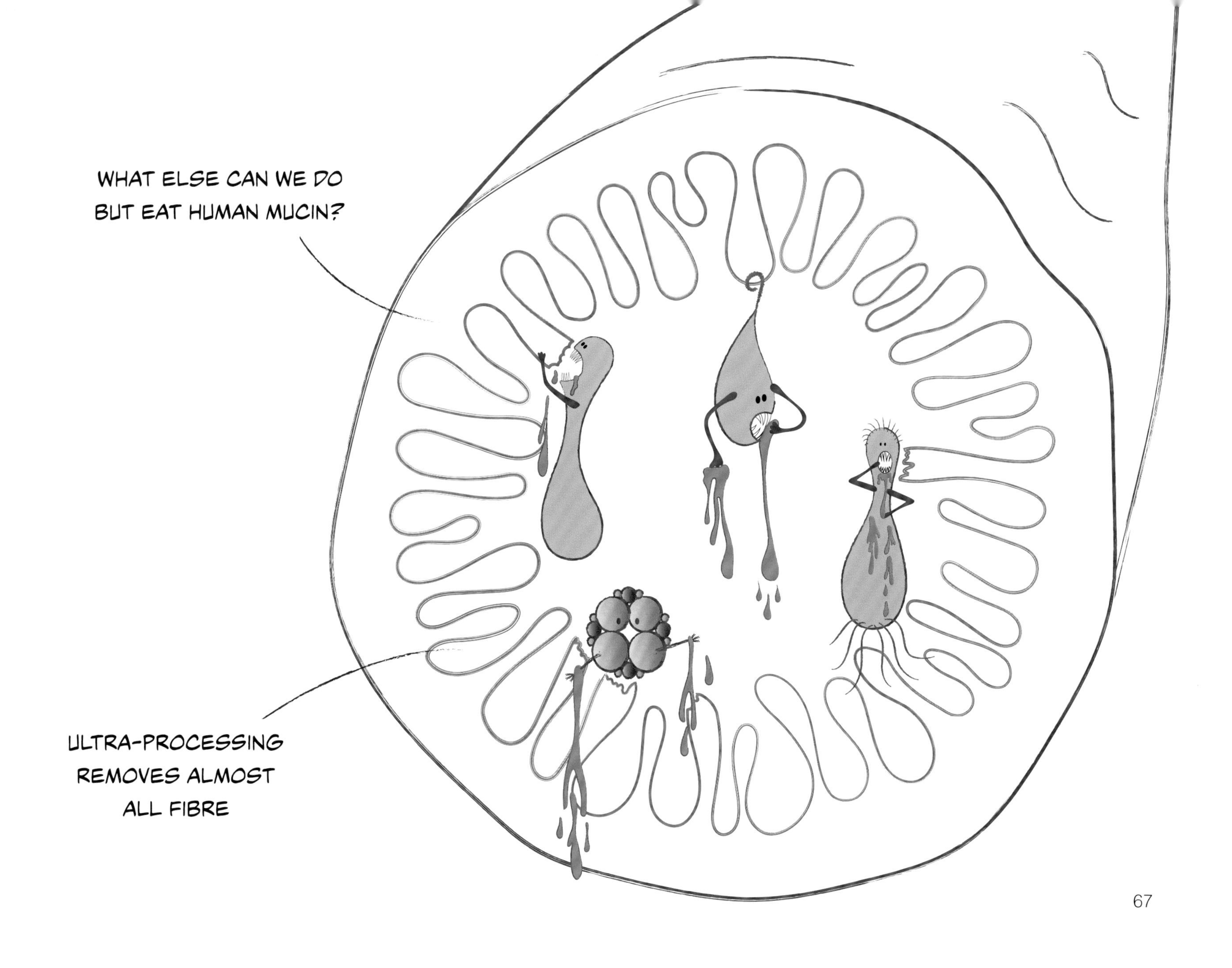
WHAT ELSE CAN WE DO BUT EAT HUMAN MUCIN?
ULTRA-PROCESSING REMOVES ALMOST ALL FIBRE

Denis Burkitt (1911-1993) was a renowned Irish surgeon who championed the use of dietary fibre to prevent many modern diseases. An easy way to check if you are eating enough fibre is to see whether your stools sink (low fibre) or float (high fibre).

WHEN DR BURKITT SHOWED THAT
LOW-FIBRE, PROCESSED FOODS ARE ASSOCIATED
WITH SMALL STOOLS AND LARGE HOSPITALS, SOME HUMANS
ACCUSED HIM OF BEING FULL OF HOT AIR!

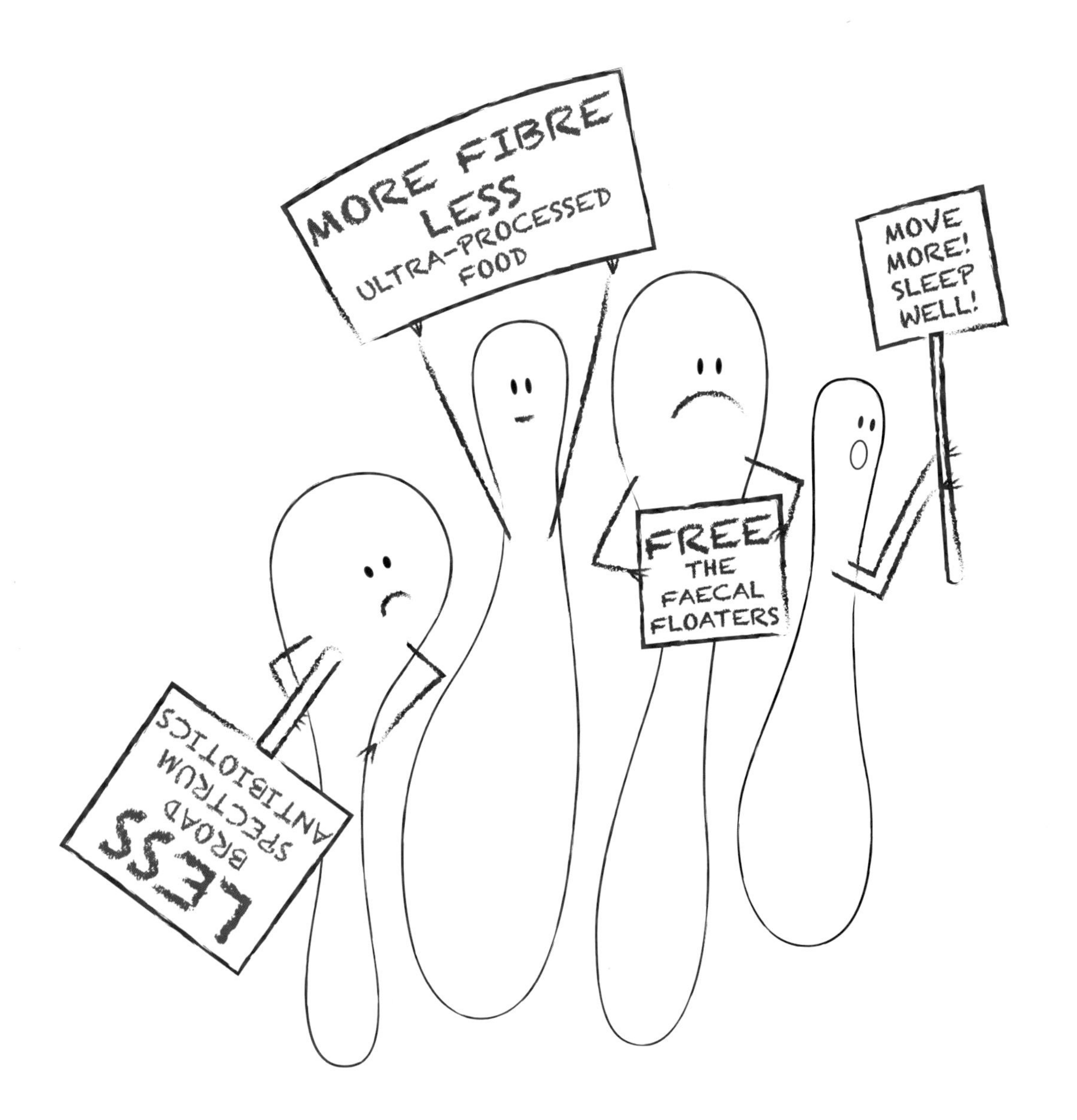
LESS BROAD SPECTRUM ANTIBIOTICS
MORE FIBRE LESS ULTRA-PROCESSED FOOD
FREE THE FAECAL FLOATERS
MOVE MORE! SLEEP WELL!

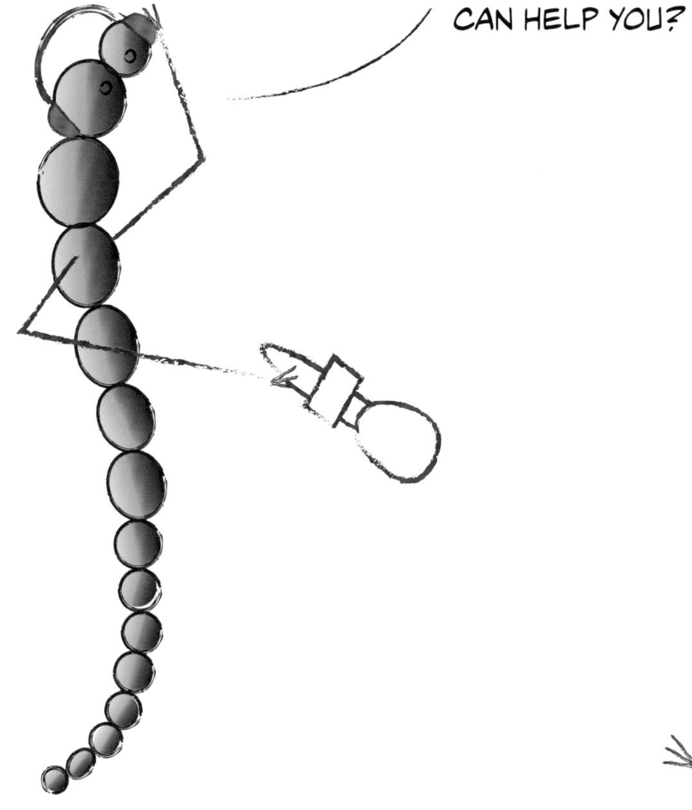
AS AN INSIDER, WHAT WOULD YOU LIKE TO TELL THE HUMANS ABOUT HOW THEY CAN HELP YOU?

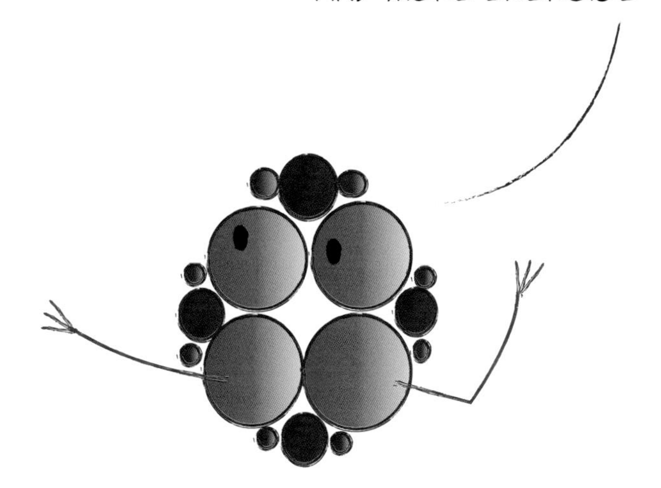
QUITE SIMPLY, IF HUMANS STOPPED ABUSING THEIR BODY, WE WOULD BE HAPPY. WE NEED BETTER FOOD, BETTER SLEEP AND MORE EXERCISE

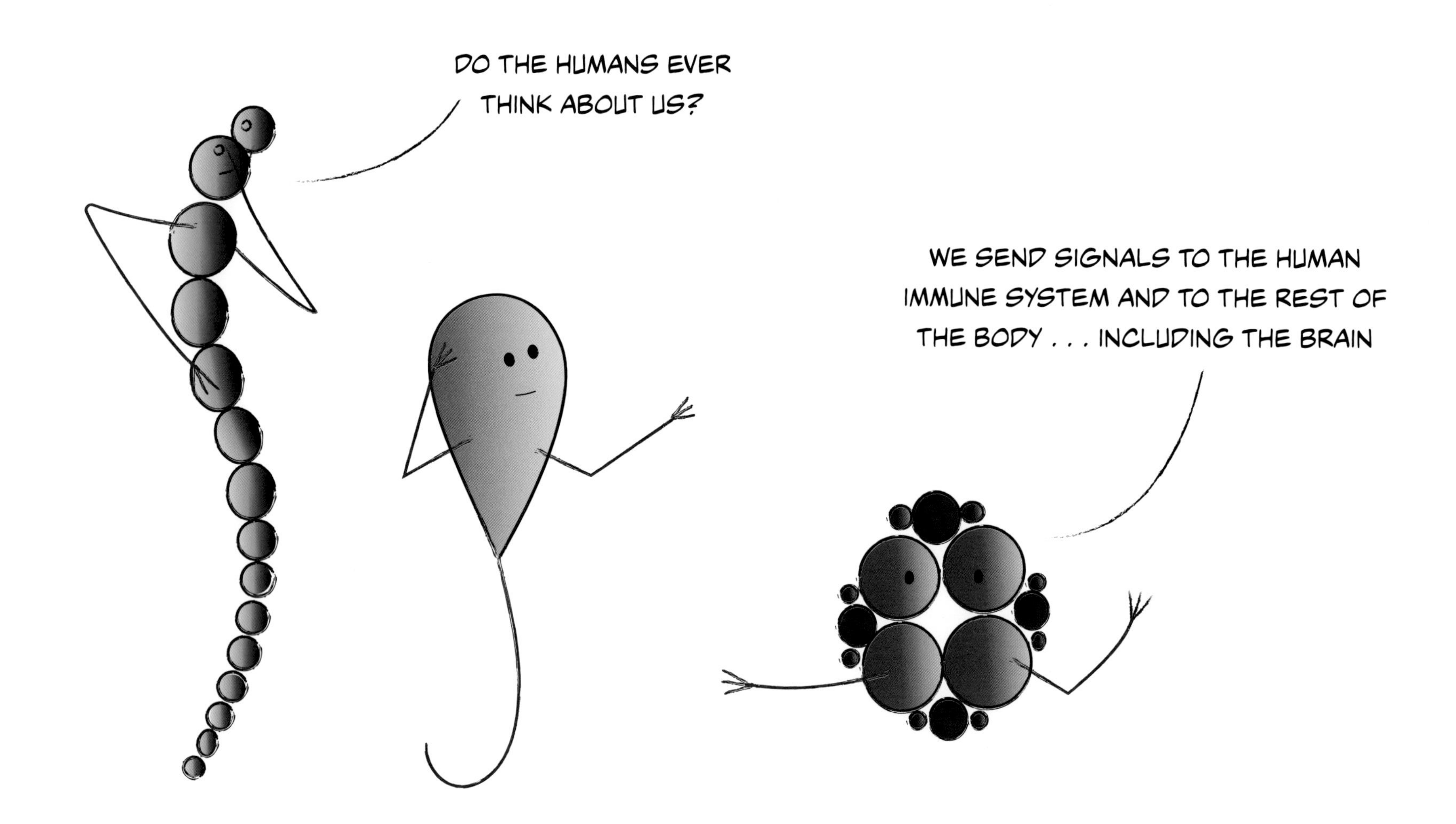
DO THE HUMANS EVER THINK ABOUT US?
WE SEND SIGNALS TO THE HUMAN IMMUNE SYSTEM AND TO THE REST OF THE BODY . . . INCLUDING THE BRAIN

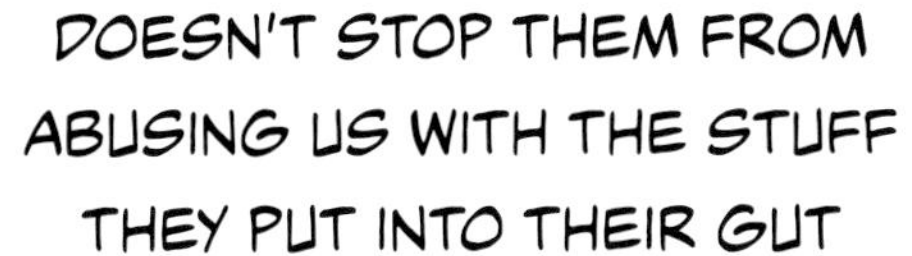
DOESN'T STOP THEM FROM ABUSING US WITH THE STUFF THEY PUT INTO THEIR GUT

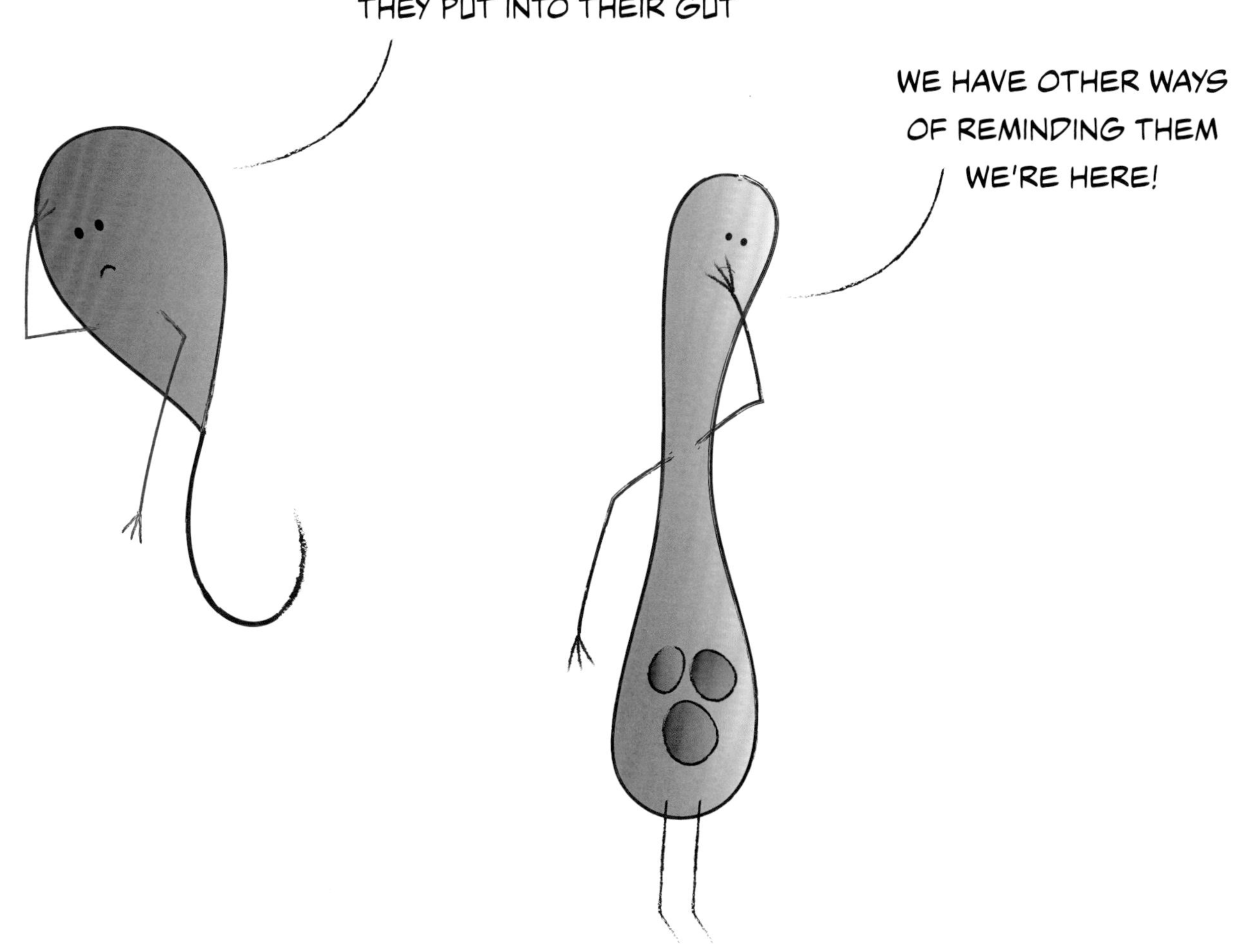
WE HAVE OTHER WAYS OF REMINDING THEM WE'RE HERE!

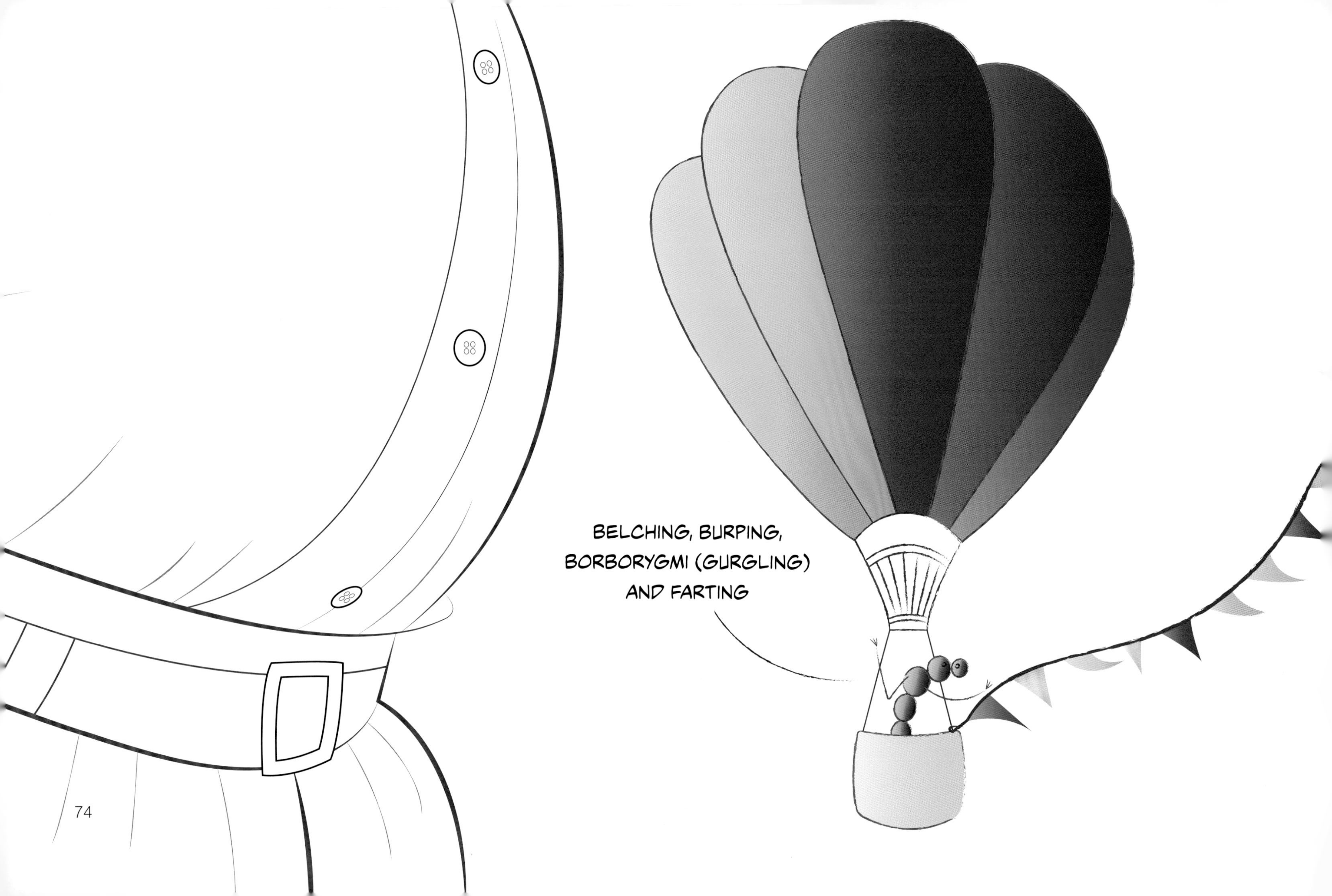
BELCHING, BURPING,
BORBORYGMI (GURGLING)
AND FARTING

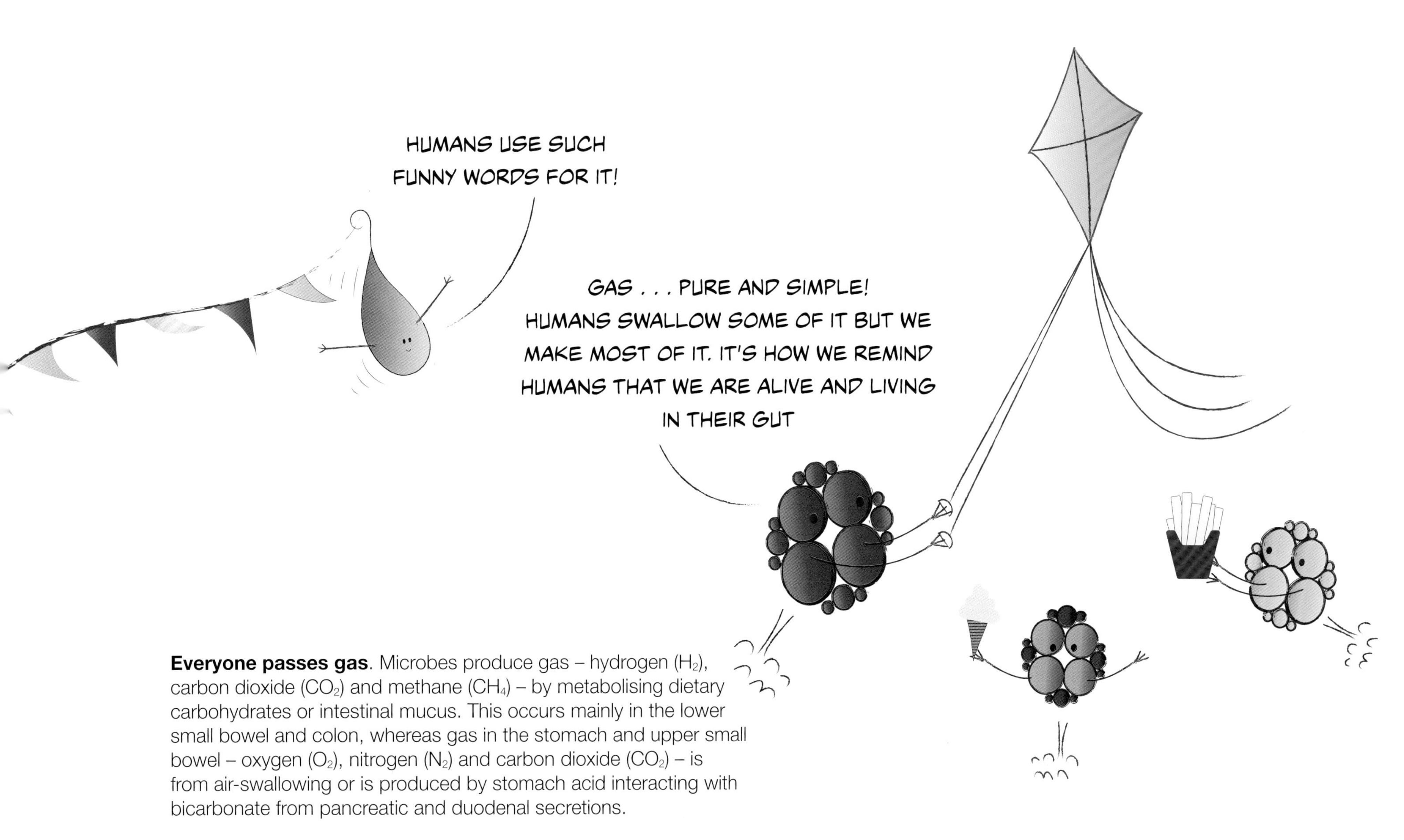

Everyone passes gas. Microbes produce gas – hydrogen (H_2), carbon dioxide (CO_2) and methane (CH_4) – by metabolising dietary carbohydrates or intestinal mucus. This occurs mainly in the lower small bowel and colon, whereas gas in the stomach and upper small bowel – oxygen (O_2), nitrogen (N_2) and carbon dioxide (CO_2) – is from air-swallowing or is produced by stomach acid interacting with bicarbonate from pancreatic and duodenal secretions.

AND ANOTHER THING;
WHEN HUMANS SMOKE . . .
WE SUFFER

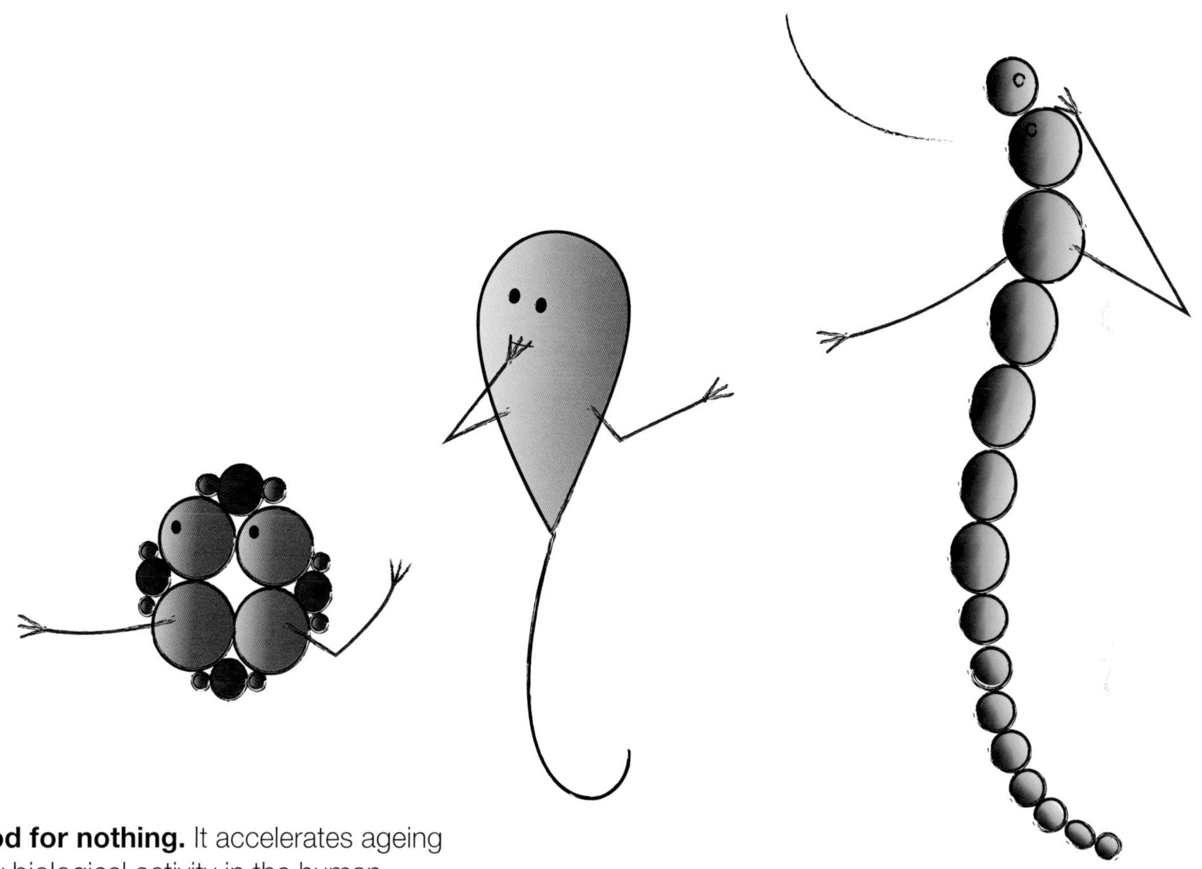

Tobacco smoking is good for nothing. It accelerates ageing and damages almost every biological activity in the human body. It also diminishes the beneficial effects of gut microbes.

Excess alcohol is toxic to microbes (as well as to humans). Although there is no such thing as a safe level of intake, toxicity is dose-dependent.

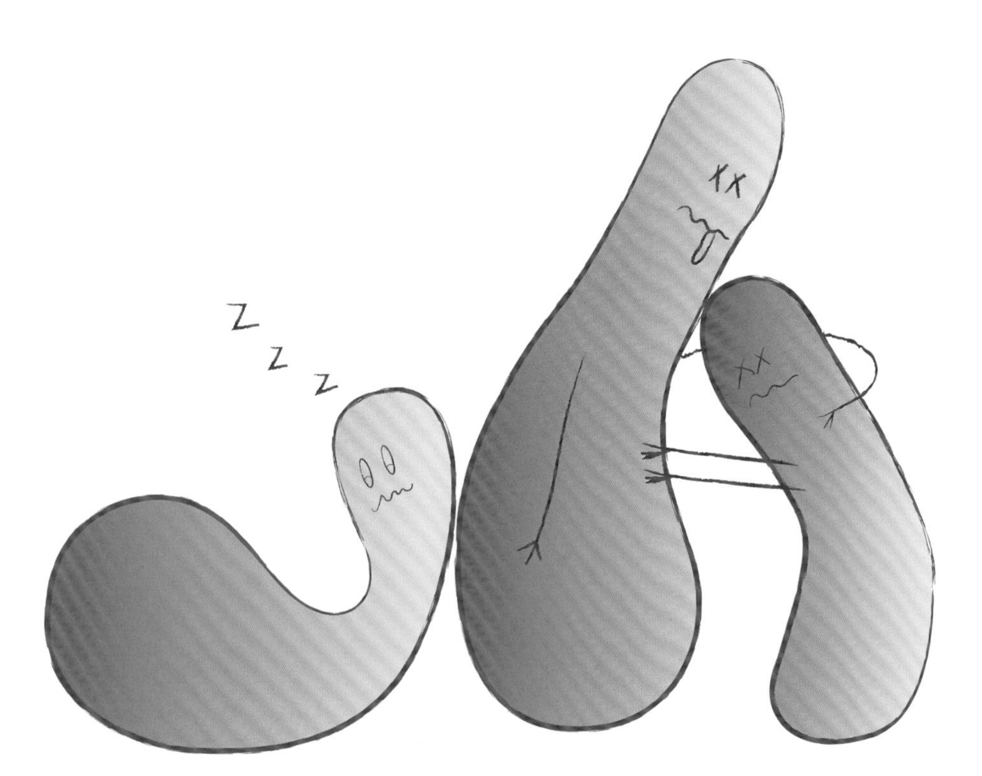

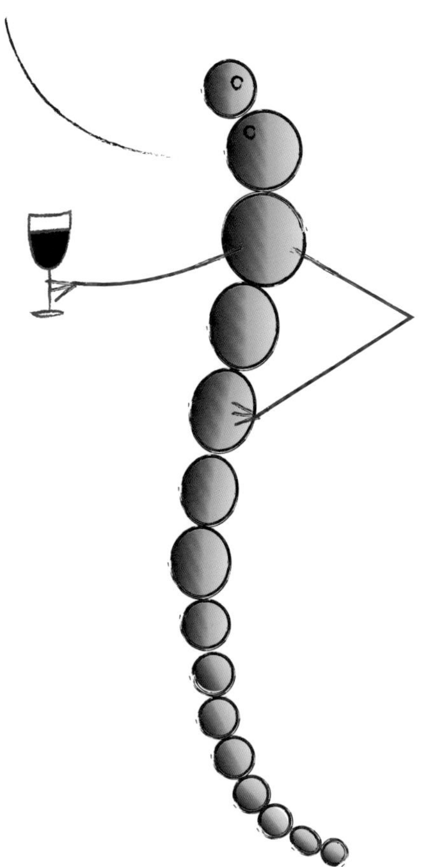

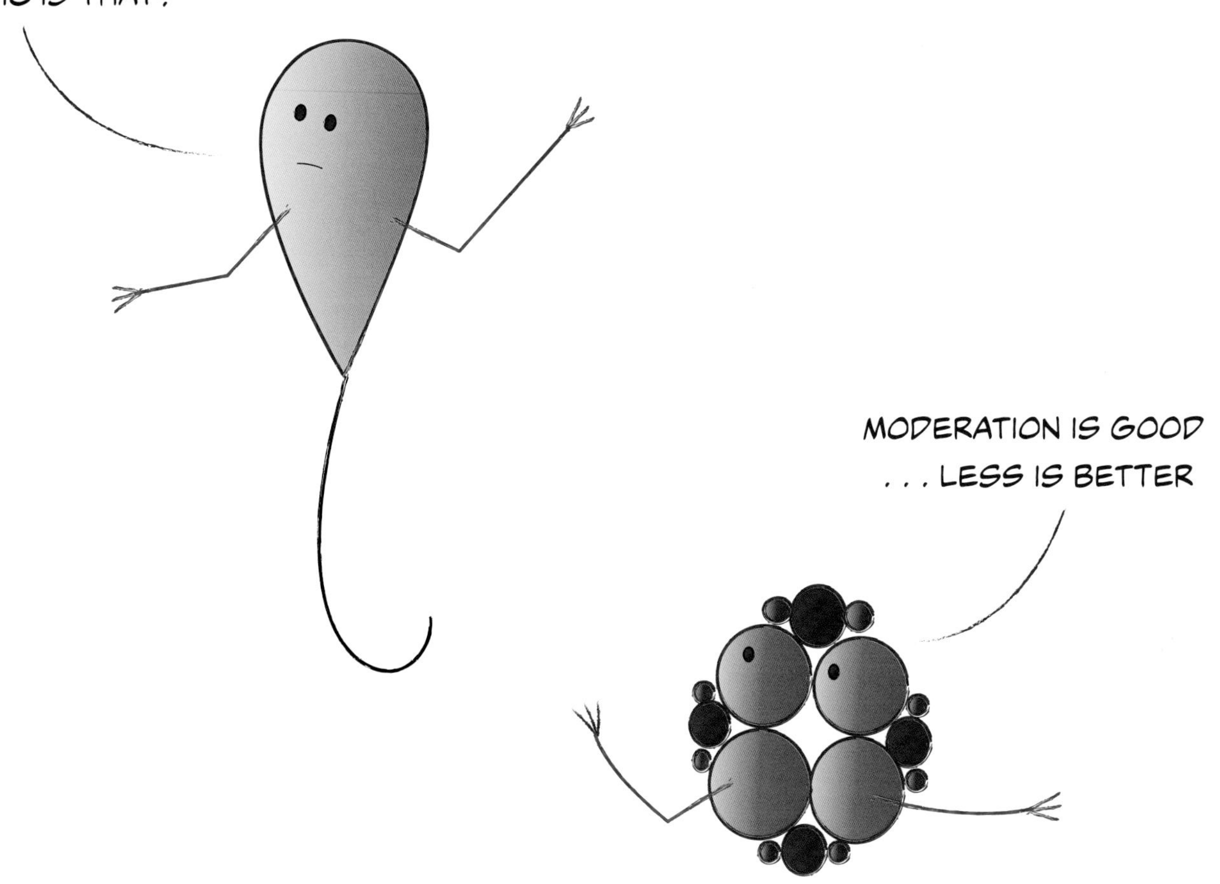
HUMANS TOASTING TO THEIR HEALTH WHILE TAKING ALCOHOL - WHAT KIND OF LOGIC IS THAT?
MODERATION IS GOOD . . . LESS IS BETTER

Regular physical activity enhances the efficiency of all body systems including human metabolism, immunity and brain function. It has also been shown that the human microbiome adapts to and improves with regular physical exercise.

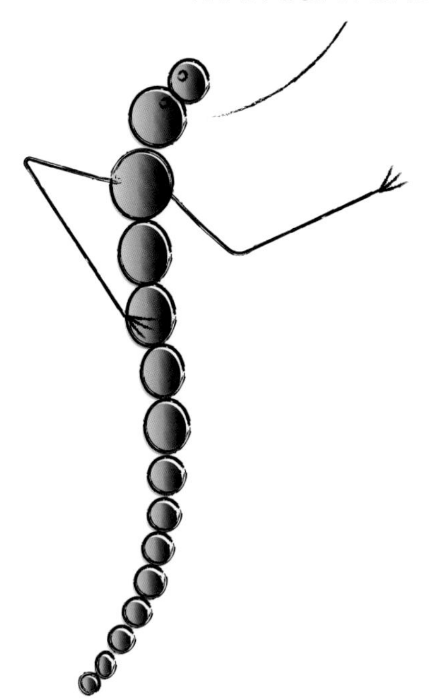

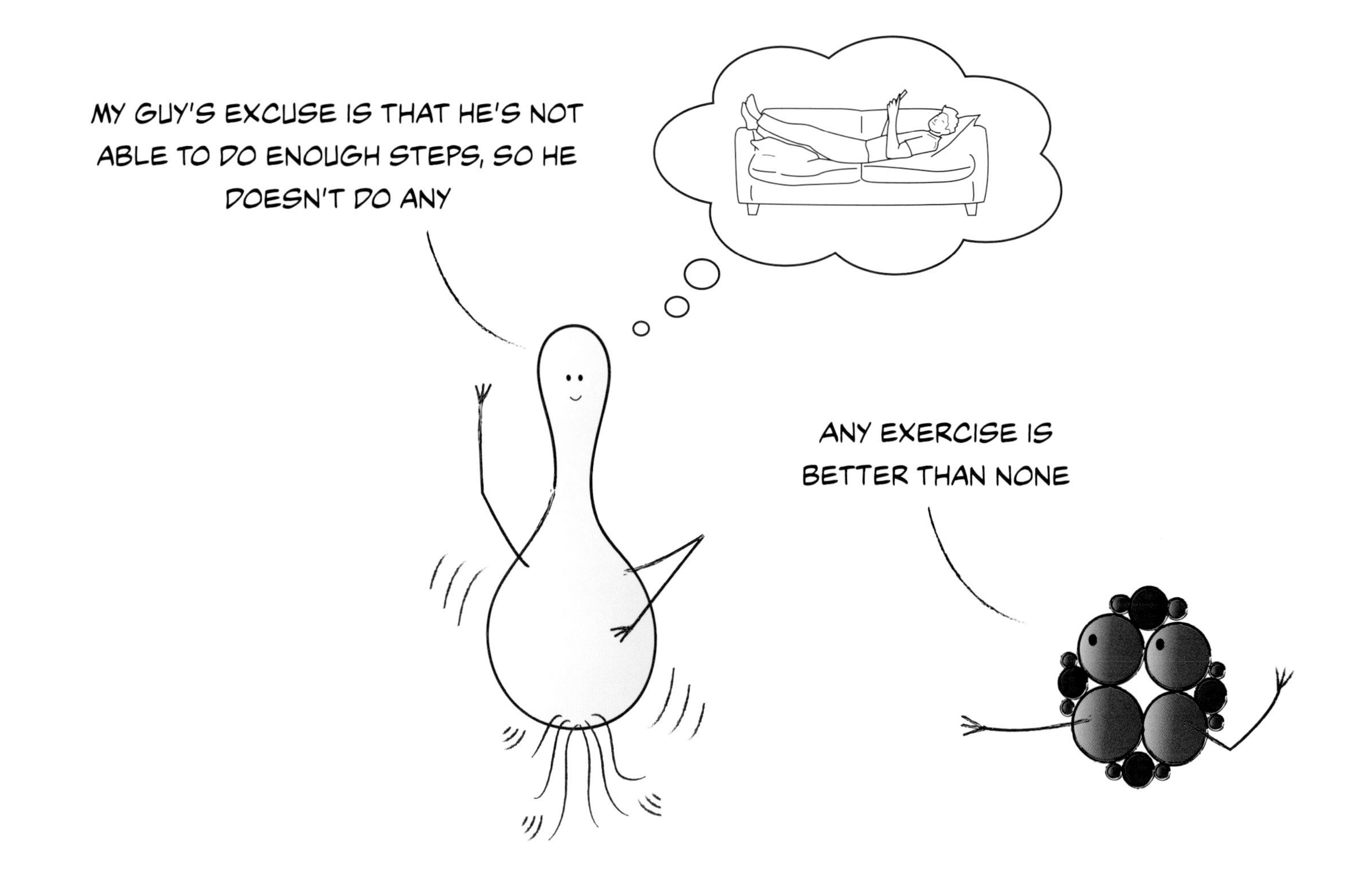
MY GUY'S EXCUSE IS THAT HE'S NOT ABLE TO DO ENOUGH STEPS, SO HE DOESN'T DO ANY
ANY EXERCISE IS BETTER THAN NONE

Probiotics are live microbes which give a health benefit if taken in sufficient amounts. This must be proven in clinical trials in humans, not in experimental animals or in a laboratory. Like microbes (and people), not all probiotics are the same. Discussing probiotics in general terms is like asking whether pills or tablets are good for you: it depends on what's in the pill and what it's intended for.

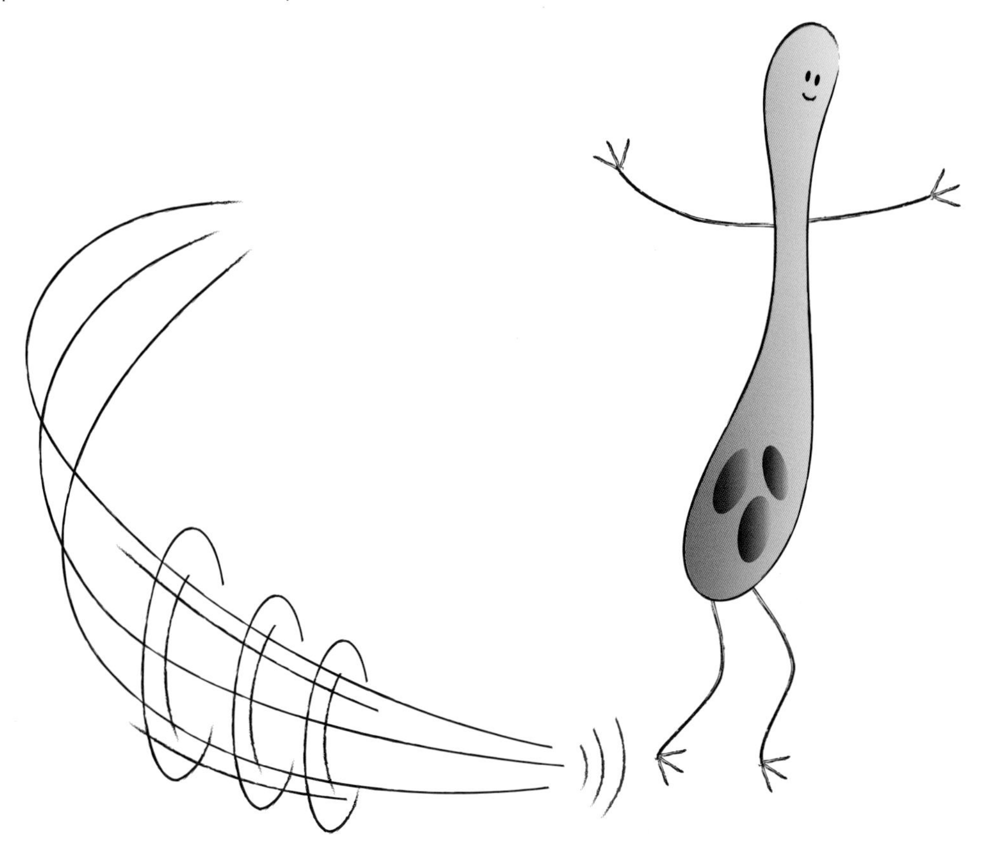

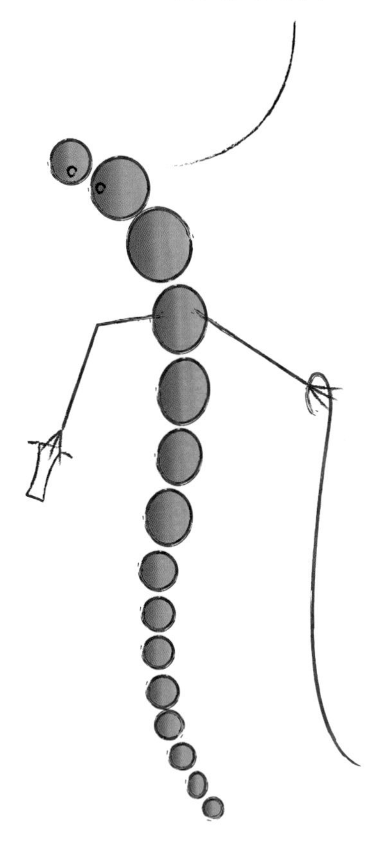

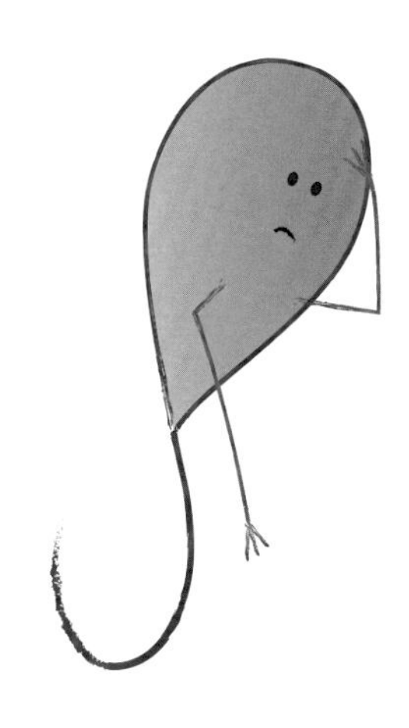

DO WE NEED SHAKING UP?

HE'S WHAT HUMANS CALL A PROBIOTIC . . . SUPPOSED TO SHAKE THINGS UP AROUND HERE!

YEAH, BUT PROBIOTICS ARE NOT ALL THE SAME!

BE SPECIFIC, ALWAYS NAME THE ORGANISM!
YOU WOULDN'T JUST TAKE A PILL OR A TABLET WITHOUT
KNOWING WHAT'S IN IT OR WHAT IT'S FOR

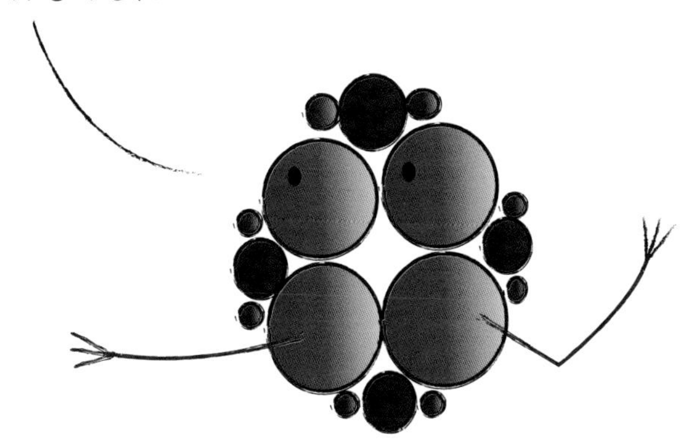

drugs 'n' bugs

'The desire to take medicine is perhaps the greatest feature which distinguishes man from animals' – the opinion of Dr William Osler, regarded as the Father of Modern Medicine. Microbes transform or inactivate commonly prescribed drugs, whereas many drugs alter the growth and function of gut microbes.

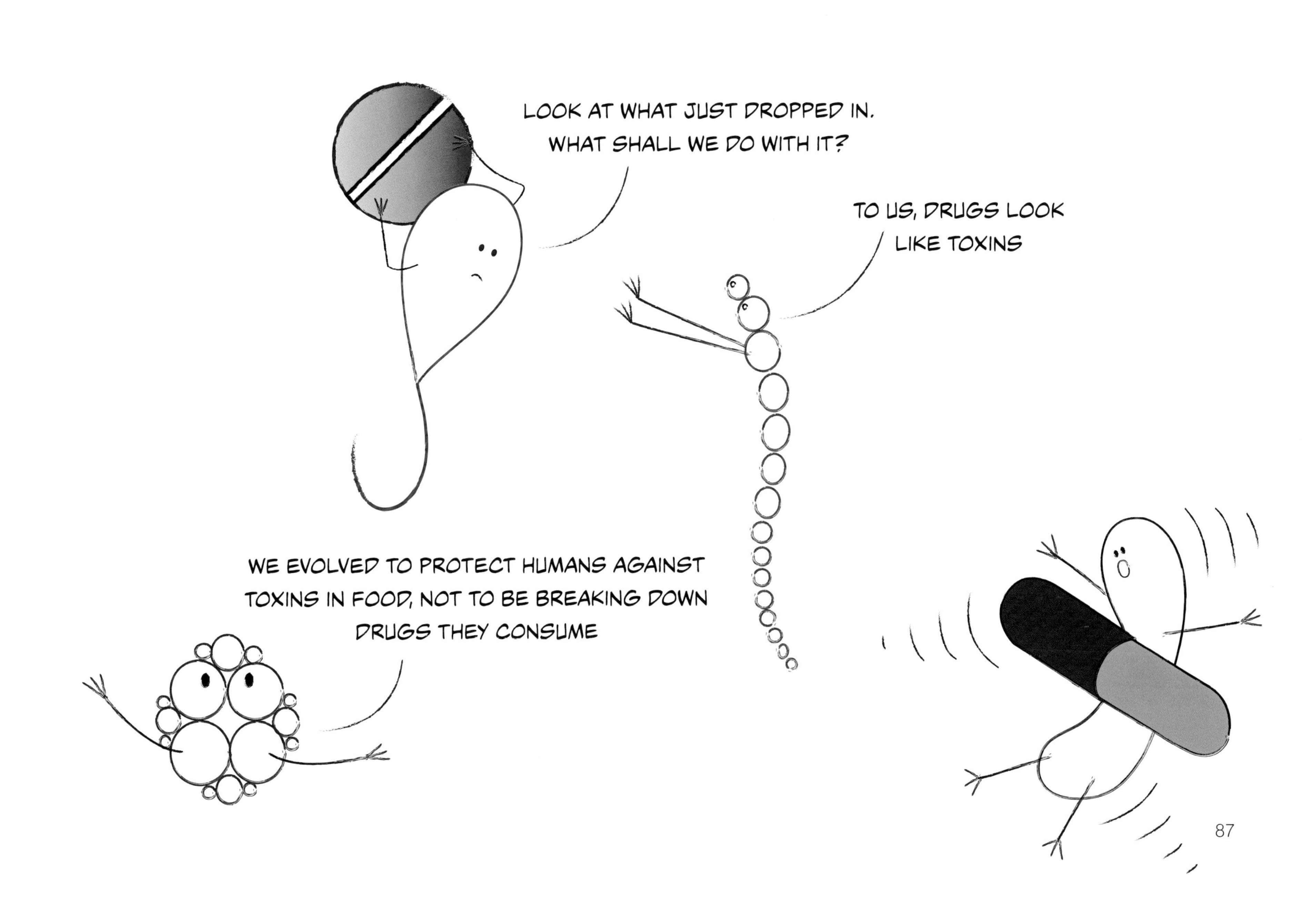
LOOK AT WHAT JUST DROPPED IN. WHAT SHALL WE DO WITH IT?
TO US, DRUGS LOOK LIKE TOXINS
WE EVOLVED TO PROTECT HUMANS AGAINST TOXINS IN FOOD, NOT TO BE BREAKING DOWN DRUGS THEY CONSUME

antimicrobial resistance

Antibiotic or antimicrobial resistance is a global crisis, spreading like a pandemic and threatening the future of healthcare. Antibiotics were actually 'invented' by ancient microbes to protect their local microenvironment long before mankind evolved. Similarly, antibiotic resistance is not new; low levels of resistance to antibiotics can be anticipated in nature because of the capacity of microbes to mutate and spread resistance genes.

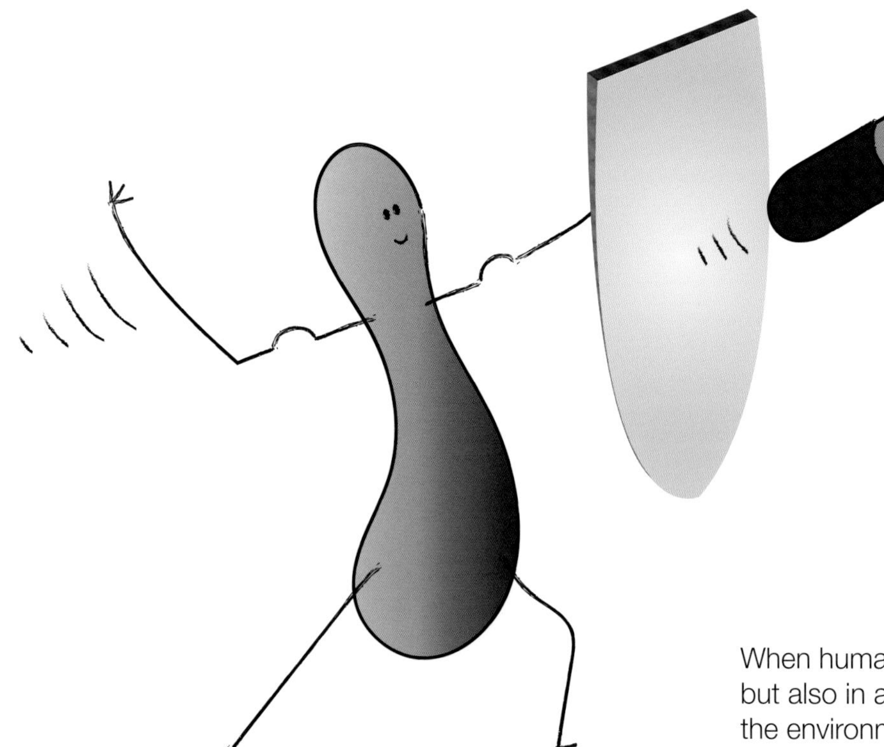

When humans 'discovered' antibiotics, intensive use not only in medicine but also in agriculture accelerated the development of resistance, polluted the environment and spread resistance across the globe.

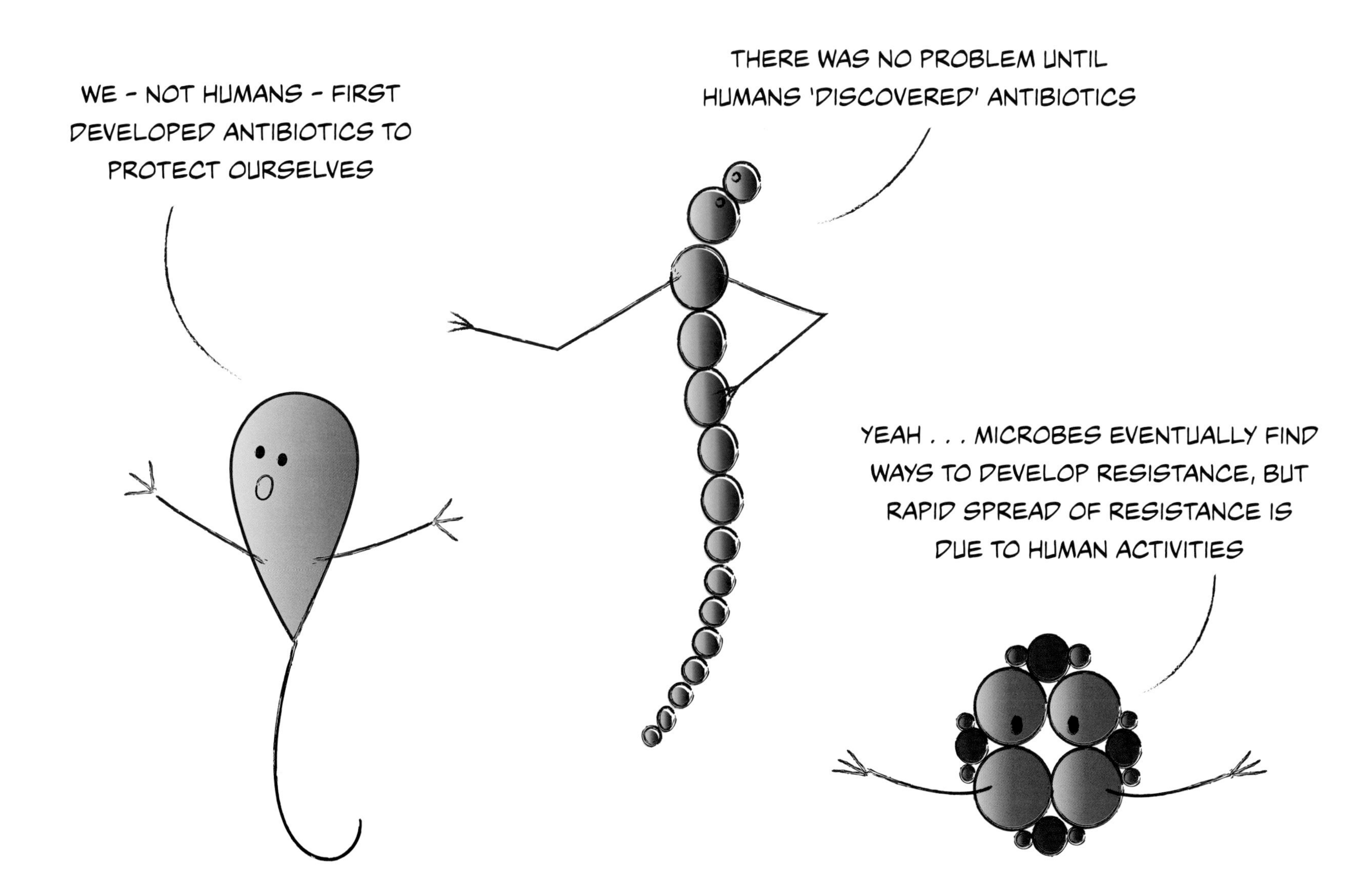
WE – NOT HUMANS – FIRST DEVELOPED ANTIBIOTICS TO PROTECT OURSELVES
THERE WAS NO PROBLEM UNTIL HUMANS 'DISCOVERED' ANTIBIOTICS
YEAH . . . MICROBES EVENTUALLY FIND WAYS TO DEVELOP RESISTANCE, BUT RAPID SPREAD OF RESISTANCE IS DUE TO HUMAN ACTIVITIES

Antimicrobial resistance is a one-world problem needing one-world cooperation to change human activities that accelerate the spread of resistance. However, public and political understanding of antimicrobial resistance is poor and often inaccurate.

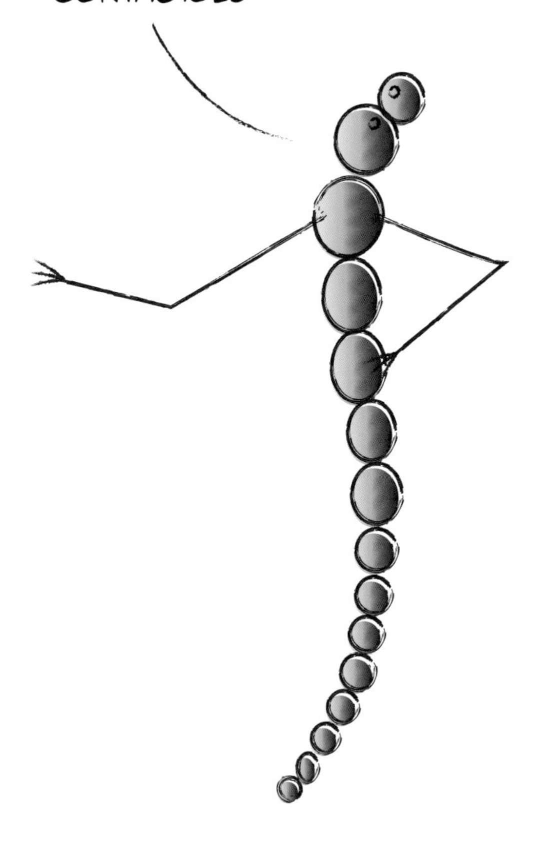
SADLY, THEY DON'T REALISE THAT RESISTANCE MAY BE CONTAGIOUS

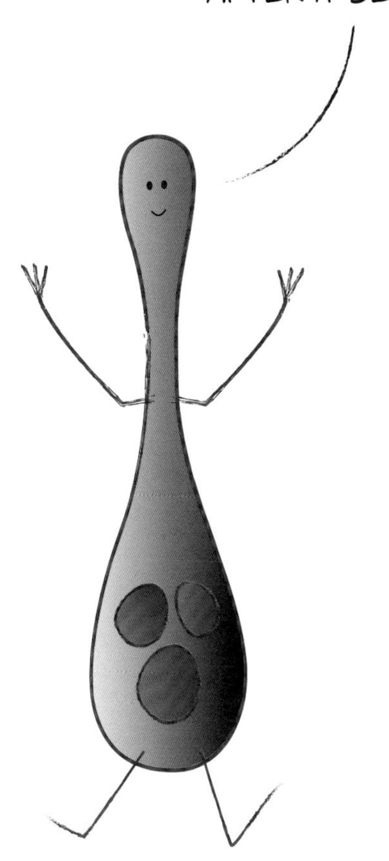
THE WORD 'SUPERBUG' REALLY BUGS ME BECAUSE IT'S THE SAME BUG AFTER IT BECOMES RESISTANT

ANTIBIOTICS HIT SOME OF US BUT LET OTHERS GROW OUT OF CONTROL

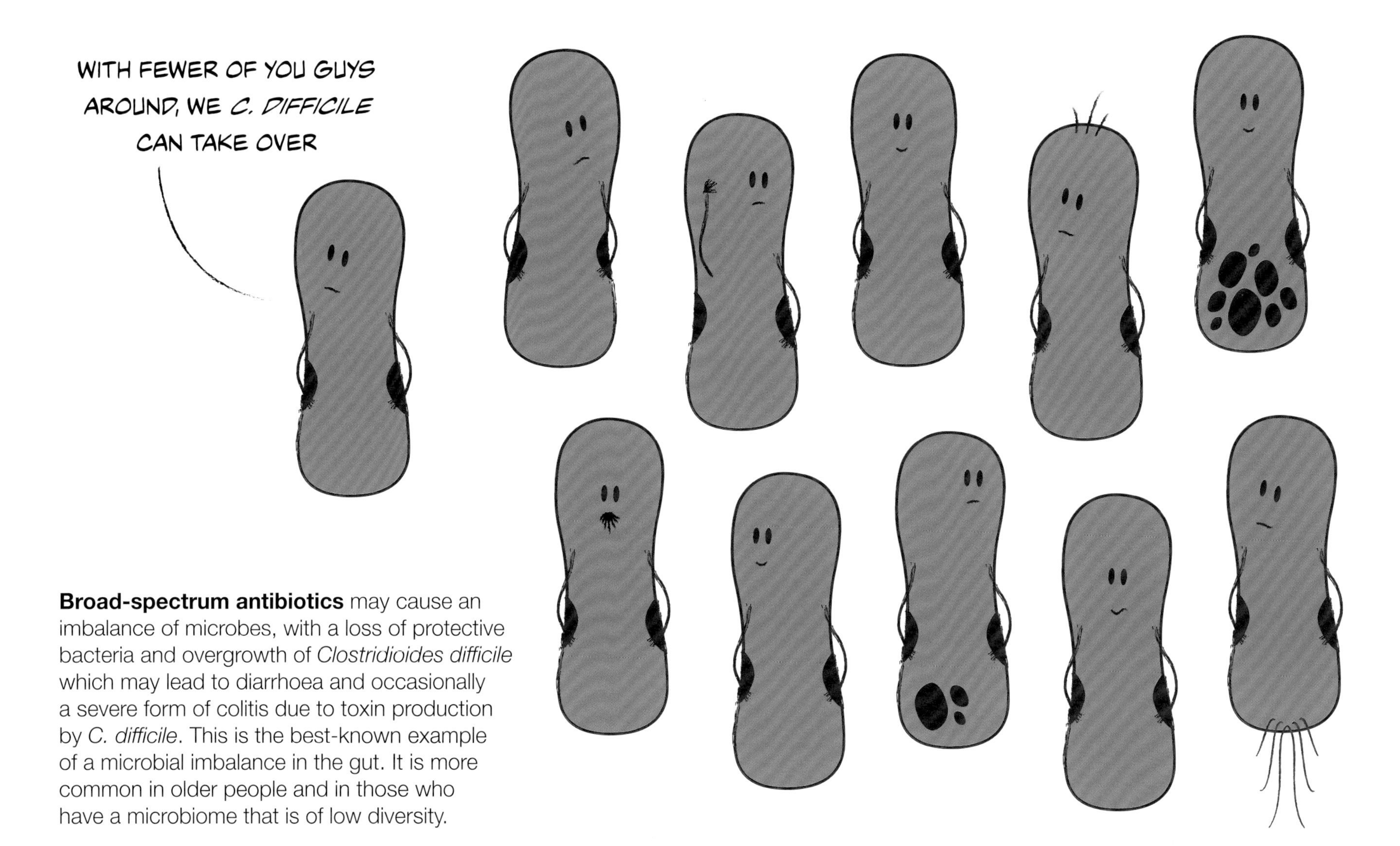

Broad-spectrum antibiotics may cause an imbalance of microbes, with a loss of protective bacteria and overgrowth of *Clostridioides difficile* which may lead to diarrhoea and occasionally a severe form of colitis due to toxin production by *C. difficile*. This is the best-known example of a microbial imbalance in the gut. It is more common in older people and in those who have a microbiome that is of low diversity.

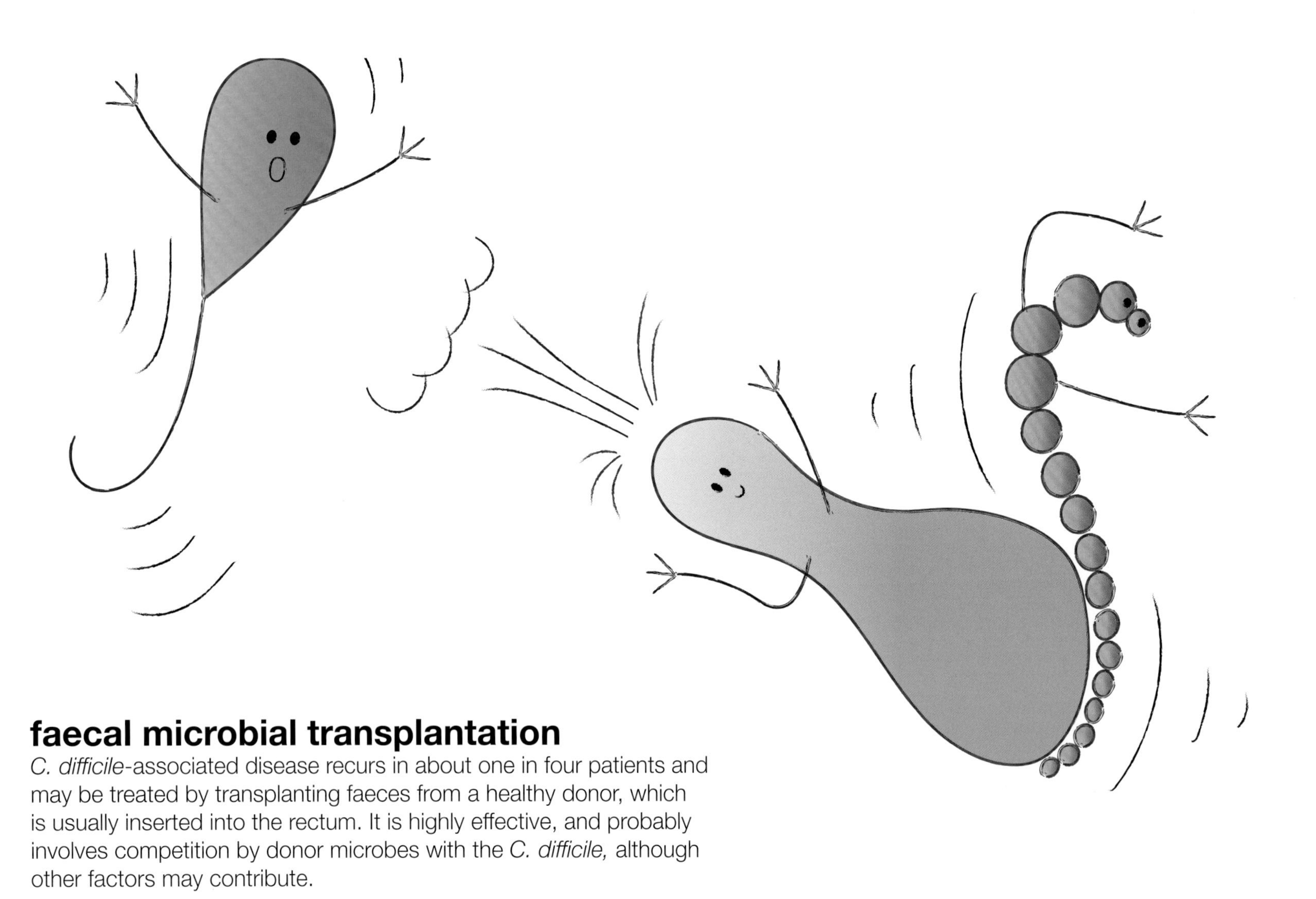

faecal microbial transplantation

C. difficile-associated disease recurs in about one in four patients and may be treated by transplanting faeces from a healthy donor, which is usually inserted into the rectum. It is highly effective, and probably involves competition by donor microbes with the *C. difficile,* although other factors may contribute.

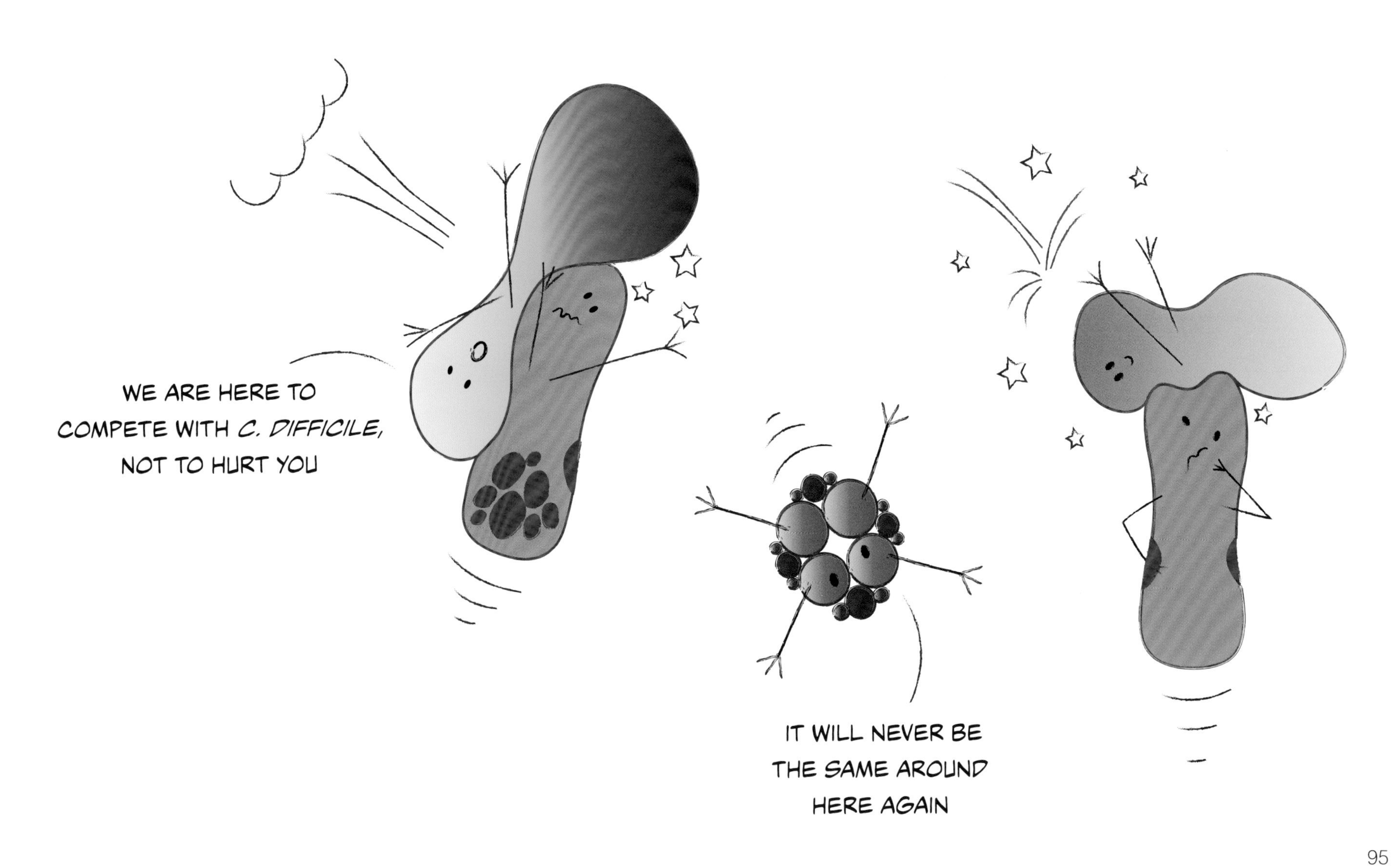
WE ARE HERE TO
COMPETE WITH *C. DIFFICILE,*
NOT TO HURT YOU
IT WILL NEVER BE
THE SAME AROUND
HERE AGAIN

phage

Phage (also known as bacteriophage) are viruses that infect bacteria. They are amongst the most common microbes on the planet and control the numbers of bacteria. They were discovered before antibiotics and represent an alternative to antibiotics for treating certain infections.

HERE'S AN ALTERNATIVE TO ANTIBIOTICS

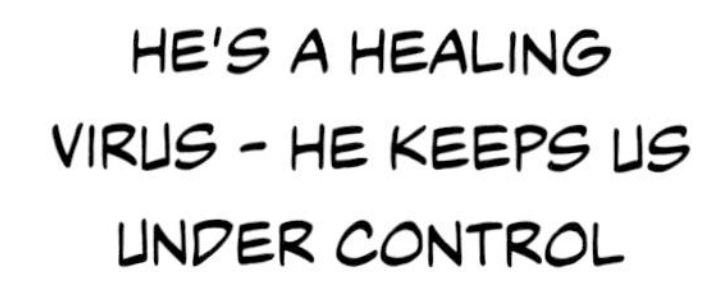

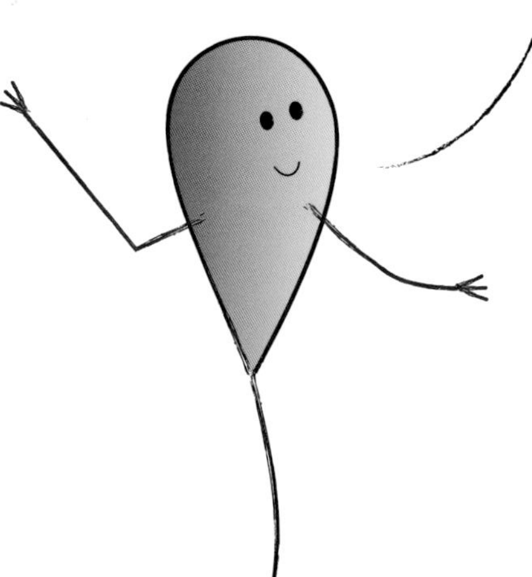

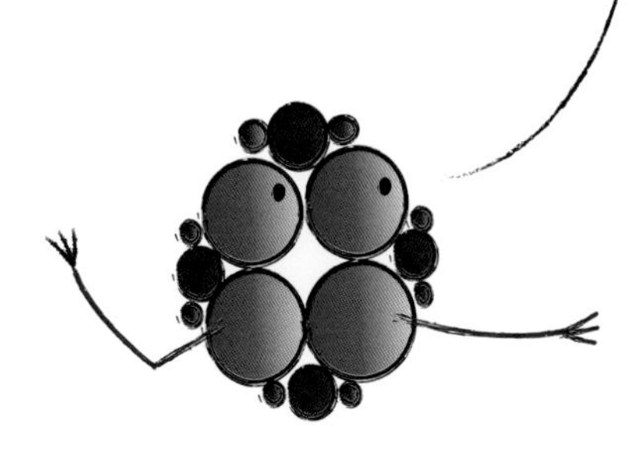

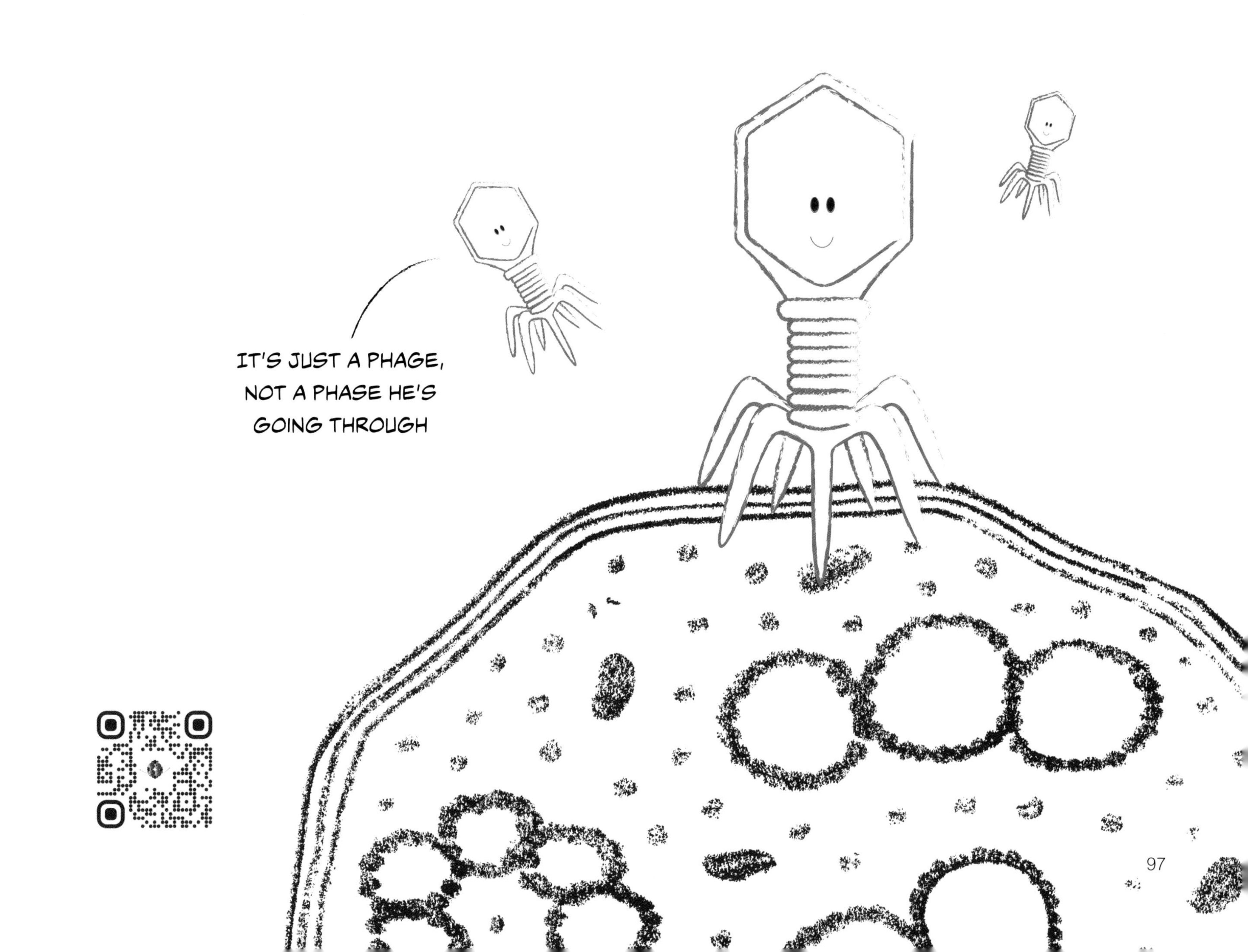
IT'S JUST A PHAGE,
NOT A PHASE HE'S
GOING THROUGH

WHEN HE GETS SICK,
WE GET BLAMED!
BUT HIS ILLNESS
AFFECTS US TOO

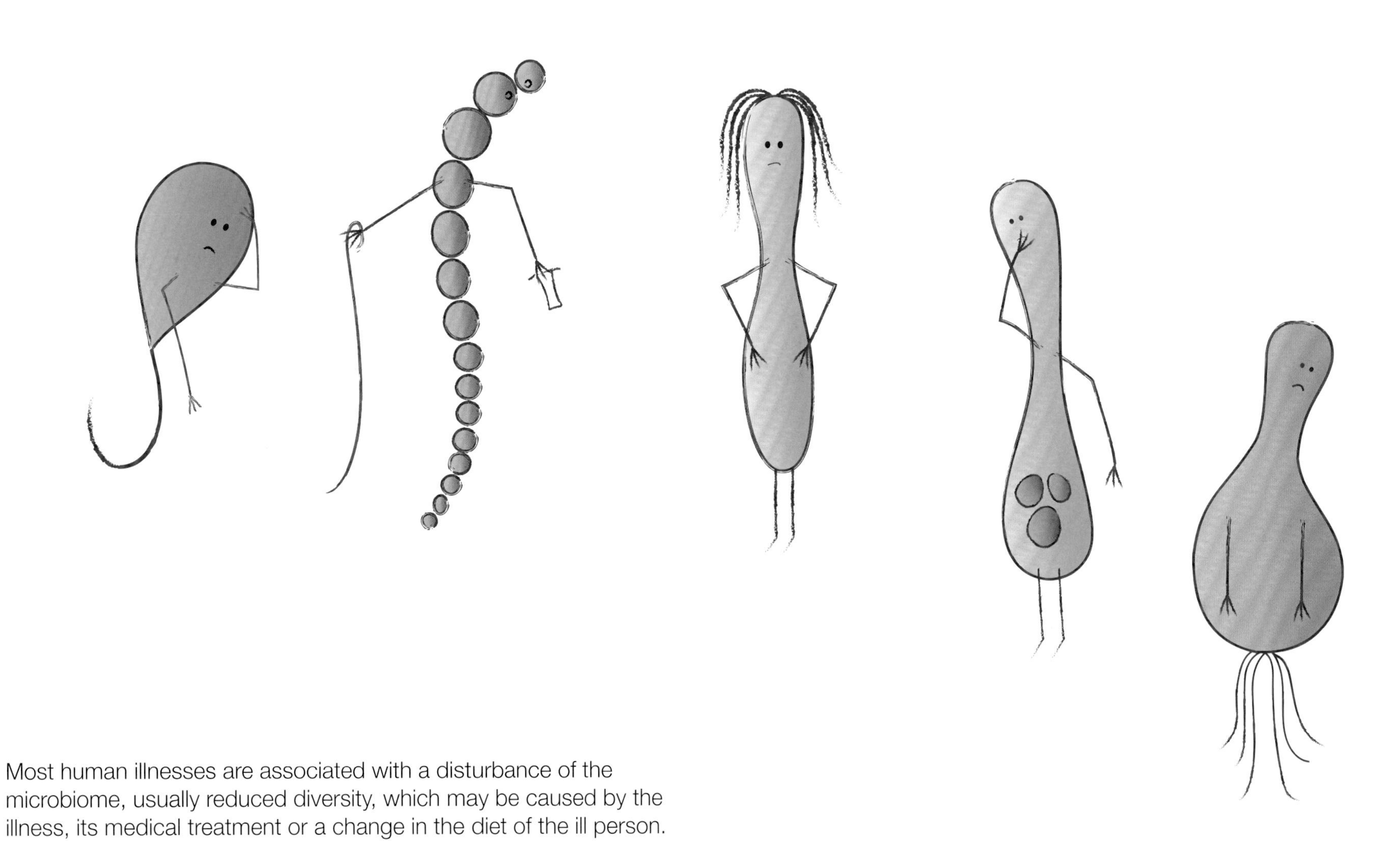

Most human illnesses are associated with a disturbance of the microbiome, usually reduced diversity, which may be caused by the illness, its medical treatment or a change in the diet of the ill person.

an obesogenic environment

Humans evolved slowly over the millennia with their personal microbes, which adapted to human lifestyle and diet. However, this human-microbe relationship may become mismatched by relatively abrupt changes in the environment and lifestyle. For example, rapid socioeconomic development and industrialisation are accompanied by the creation of an obesogenic environment (readily available energy-dense foods and reduced requirement for physical activity). The problem is increased by exposure to antibiotics, which change the microbiome and may have led to silent extinctions of many microbial species. Other modern lifestyles which have modified the microbiome include how humans are born (naturally or by C-section), how infants are fed (breastmilk or formula), the diet we consume and the environment we live in (urban or rural). Negative effects on the microbiome are cumulative over time and across human generations.

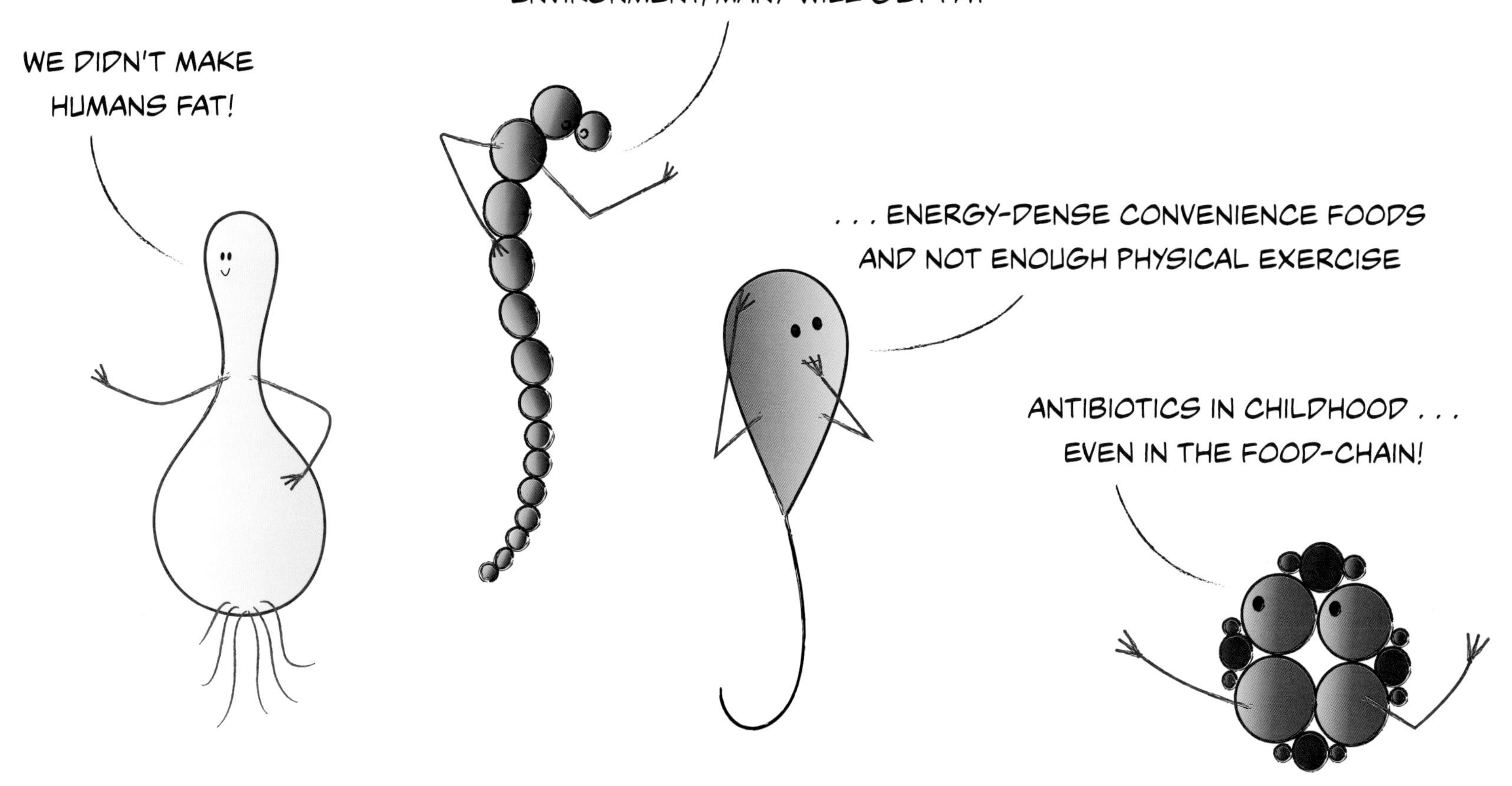
WE DIDN'T MAKE HUMANS FAT!
IT'S NOT THEIR FAULT EITHER. AS LONG AS THEY LIVE IN AN OBESOGENIC ENVIRONMENT, MANY WILL GET FAT
. . . ENERGY-DENSE CONVENIENCE FOODS AND NOT ENOUGH PHYSICAL EXERCISE
ANTIBIOTICS IN CHILDHOOD . . . EVEN IN THE FOOD-CHAIN!

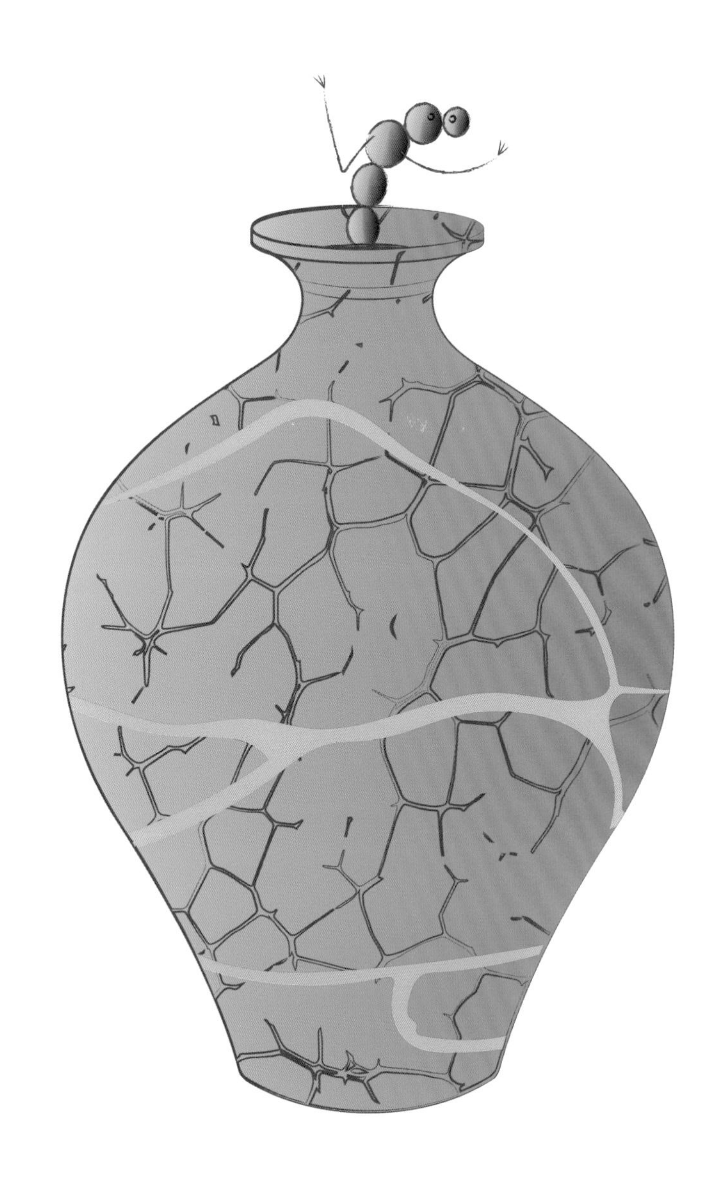

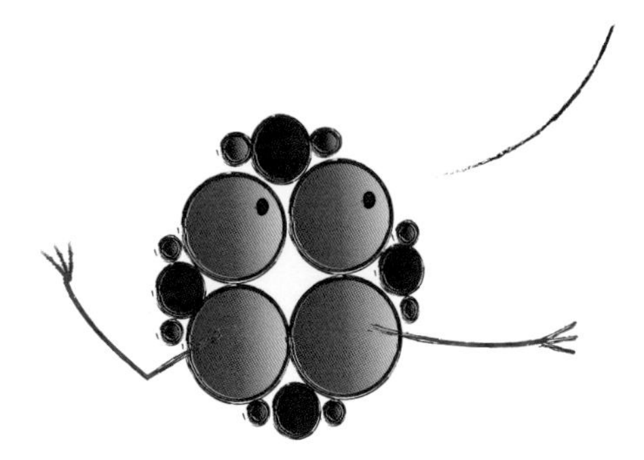
WE ARE ALL A BIT BROKEN - BUT NEVER BROKEN IN EXACTLY THE SAME WAY

broken host or broken biome

Although disturbances of the microbiome have been observed in many human disorders, in most instances it is unclear when these are either the cause, the consequence, or a contributory modifier of the disease.

what's love got to do with it?

Inspired by the opening lines of Leo Tolstoy's *Anna Karenina*, some scientists believe that healthy microbiomes (like happy families) are all alike but every unhealthy microbiome (like every unhappy family) is unhappy in its own way.

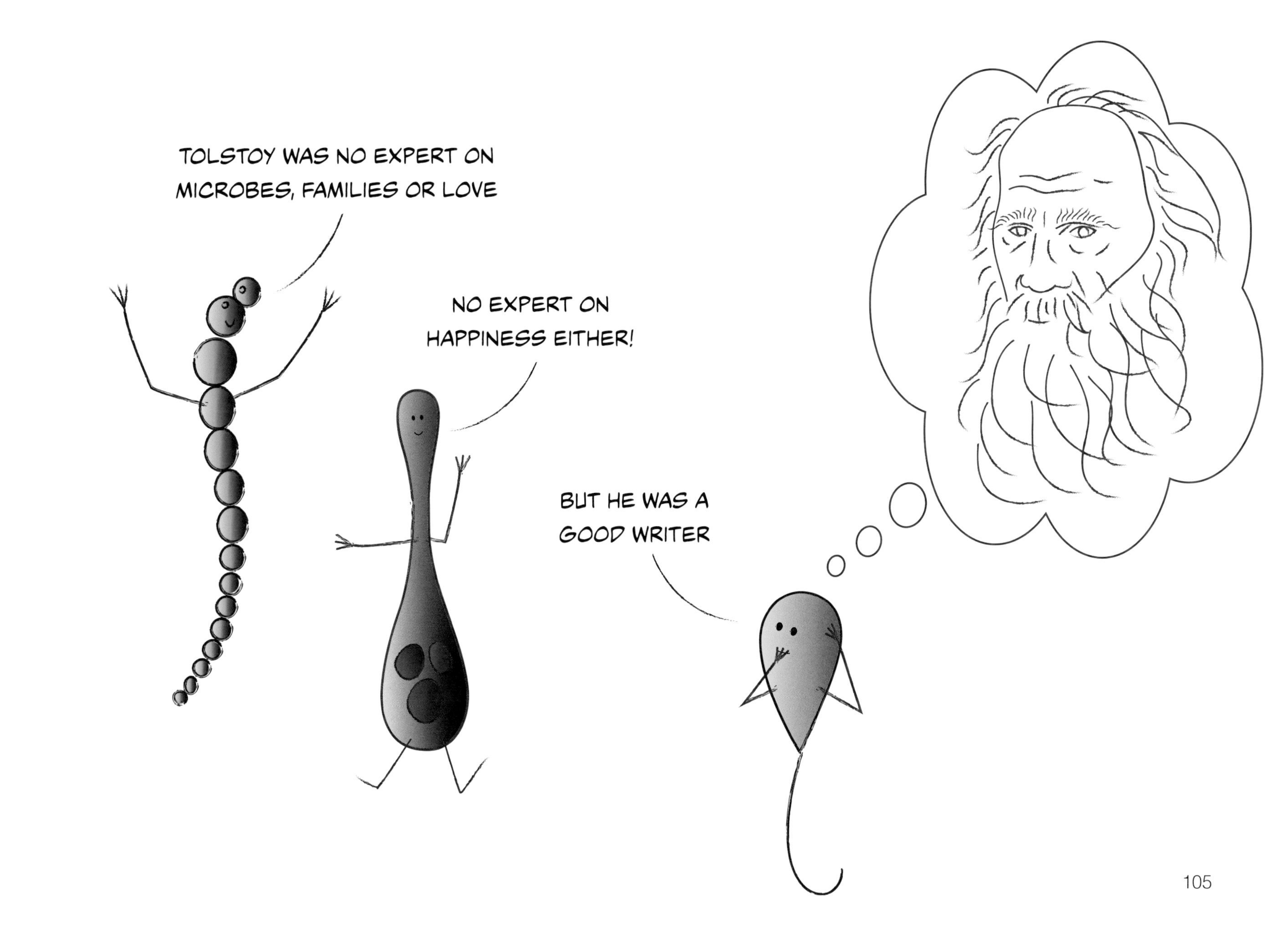
TOLSTOY WAS NO EXPERT ON MICROBES, FAMILIES OR LOVE
NO EXPERT ON HAPPINESS EITHER!
BUT HE WAS A GOOD WRITER

unusual microbiomes

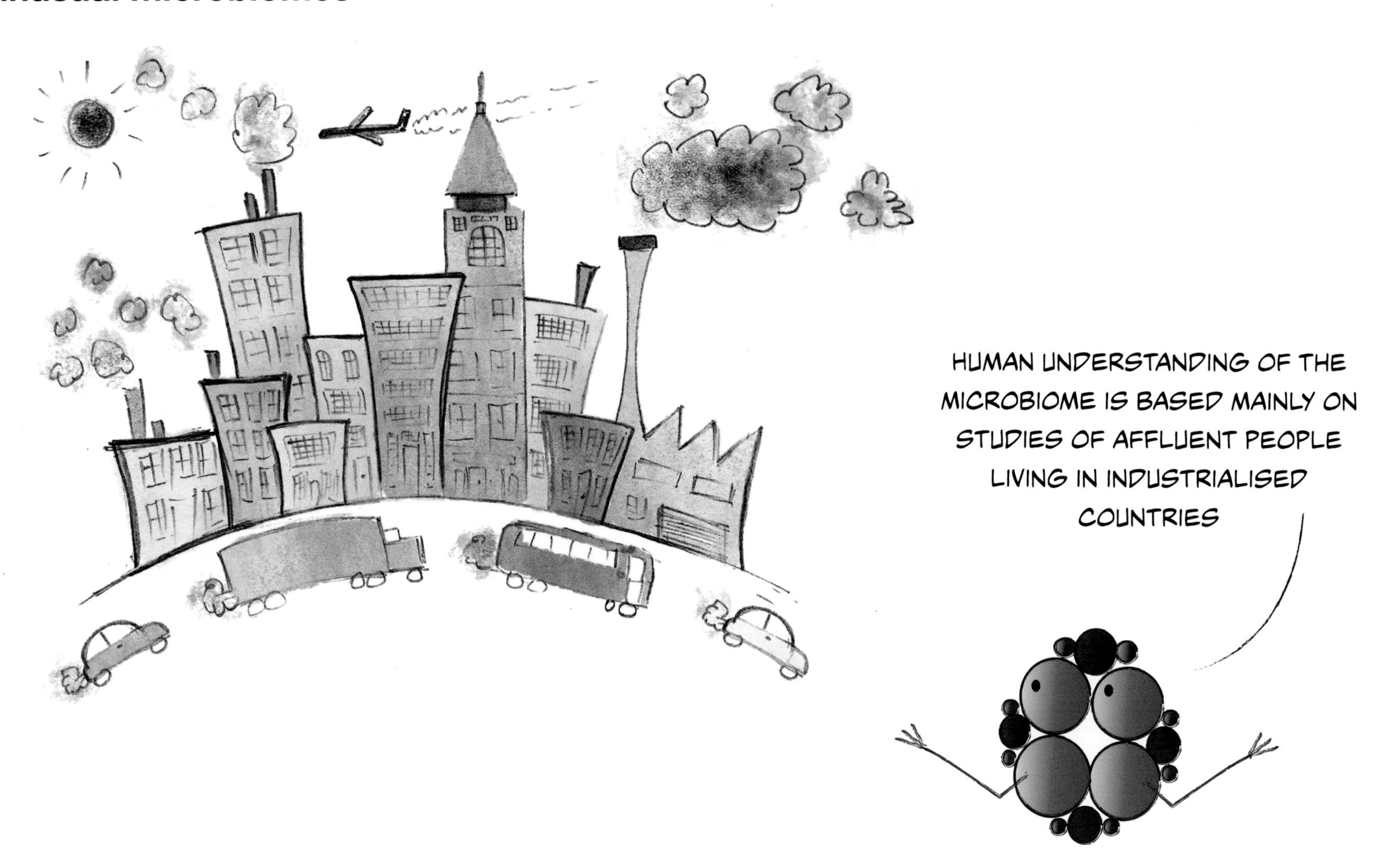

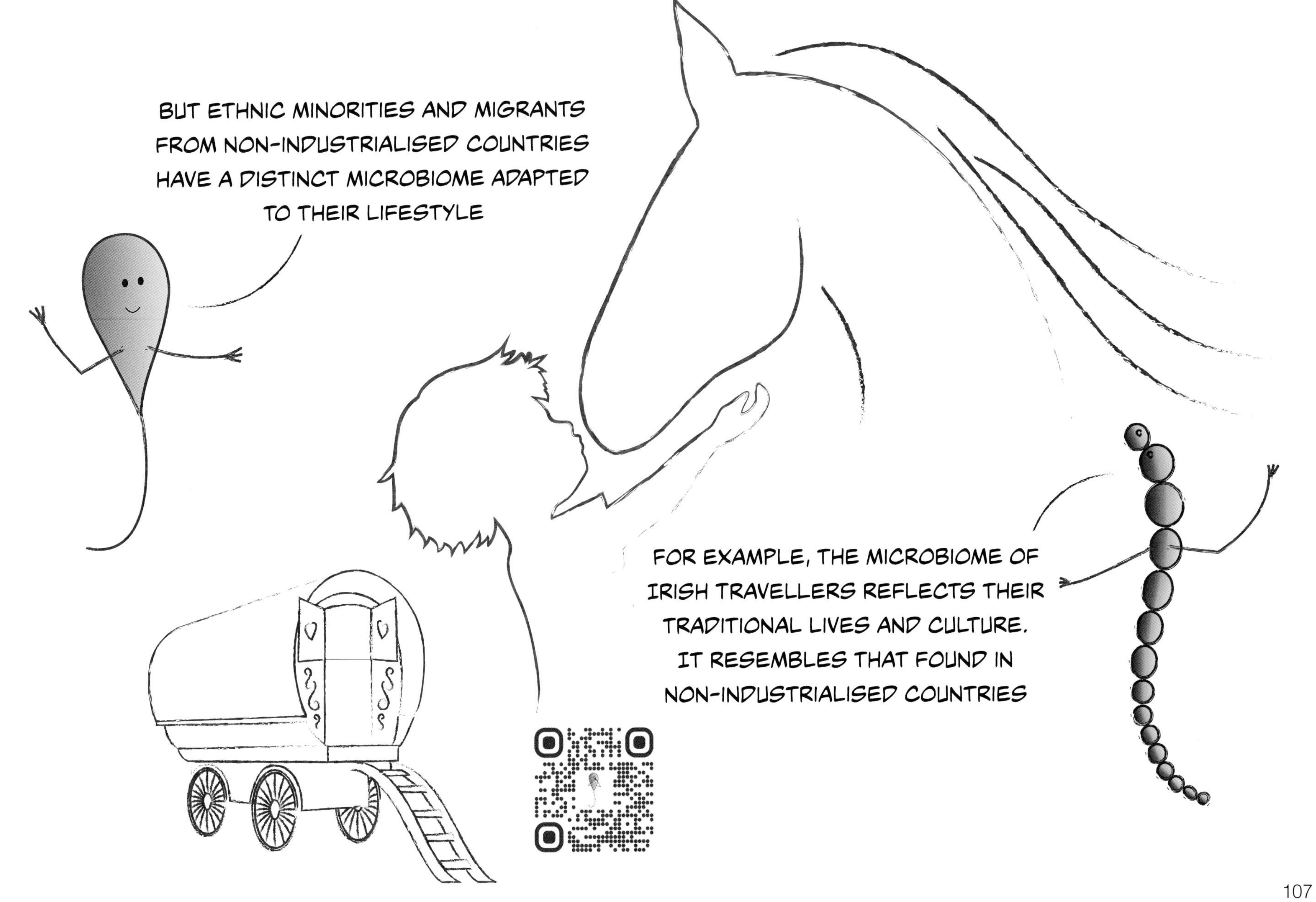
BUT ETHNIC MINORITIES AND MIGRANTS FROM NON-INDUSTRIALISED COUNTRIES HAVE A DISTINCT MICROBIOME ADAPTED TO THEIR LIFESTYLE
FOR EXAMPLE, THE MICROBIOME OF IRISH TRAVELLERS REFLECTS THEIR TRADITIONAL LIVES AND CULTURE. IT RESEMBLES THAT FOUND IN NON-INDUSTRIALISED COUNTRIES

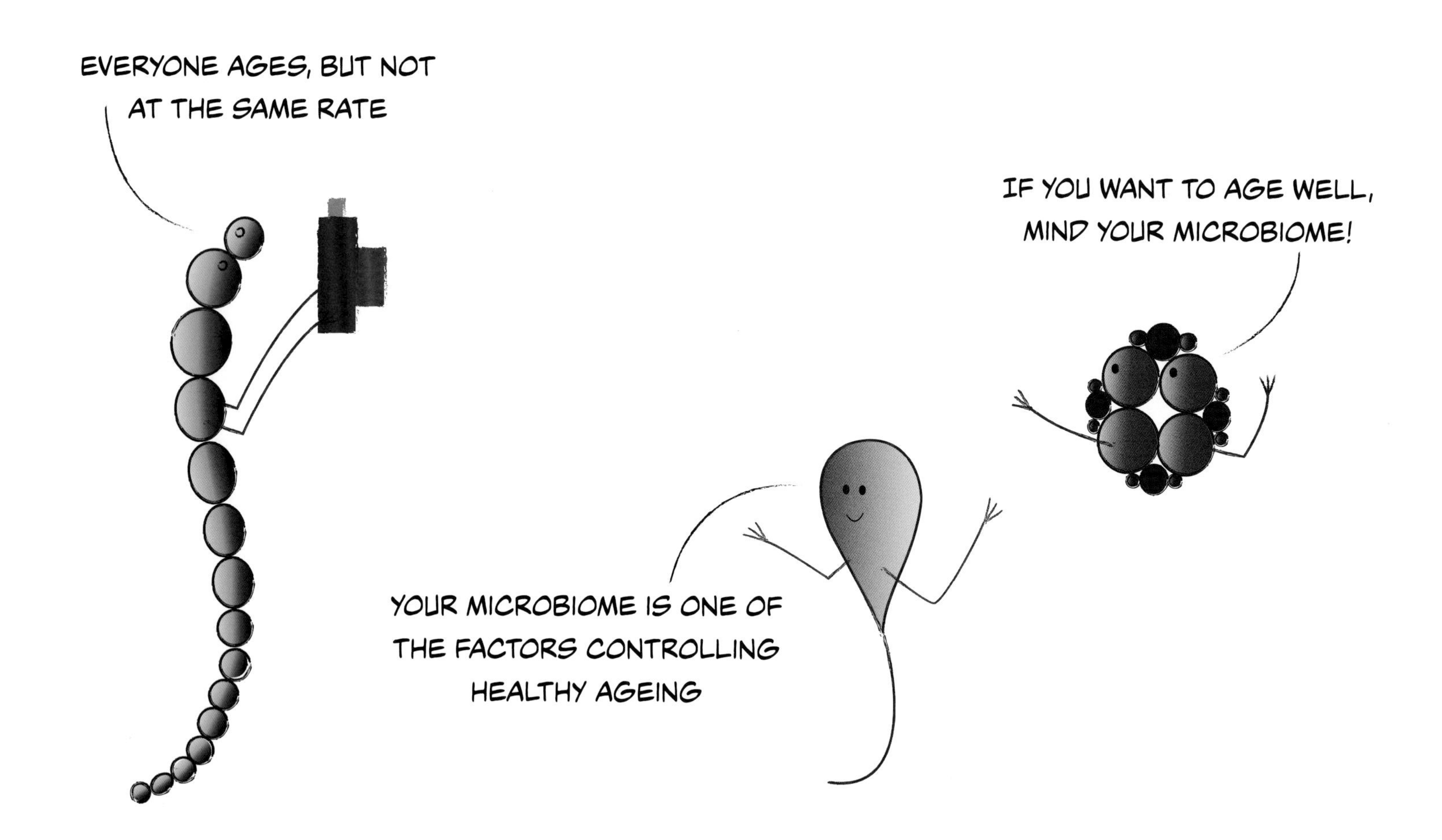
EVERYONE AGES, BUT NOT AT THE SAME RATE
YOUR MICROBIOME IS ONE OF THE FACTORS CONTROLLING HEALTHY AGEING
IF YOU WANT TO AGE WELL, MIND YOUR MICROBIOME!

growing old together

The microbiome has a complex relationship with ageing. While the microbiome affects the rate at which people age, ageing and age-related diseases alter the microbiome.

After the microbiome has been assembled during the first few years of life, there is a period of relative stability during adolescence and early adulthood, but sadly as humans age their microbiome deteriorates, with the loss of those microbes that are especially protective. This is due to age-related changes in gut physiology, loss of dietary diversity and other behavioural changes.

Although everyone ages, ageing is not experienced uniformly. People undergo age-related health loss at different rates, and the microbiome may be used as a diagnostic marker of healthy and unhealthy ageing.

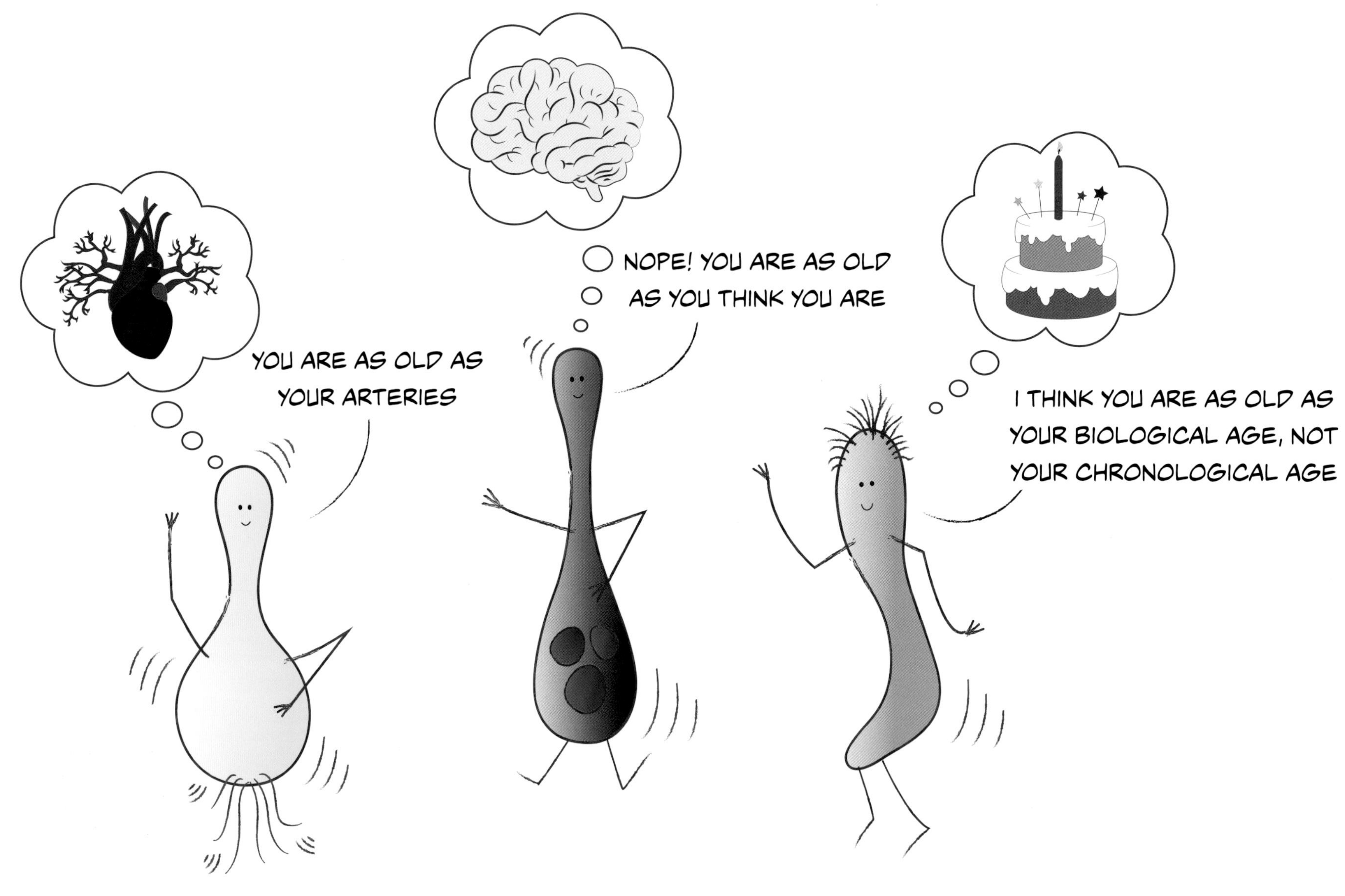
YOU ARE AS OLD AS
YOUR ARTERIES
NOPE! YOU ARE AS OLD
AS YOU THINK YOU ARE
I THINK YOU ARE AS OLD AS
YOUR BIOLOGICAL AGE, NOT
YOUR CHRONOLOGICAL AGE

ACTUALLY, YOU ARE AS OLD
AS YOUR MICROBIOME!

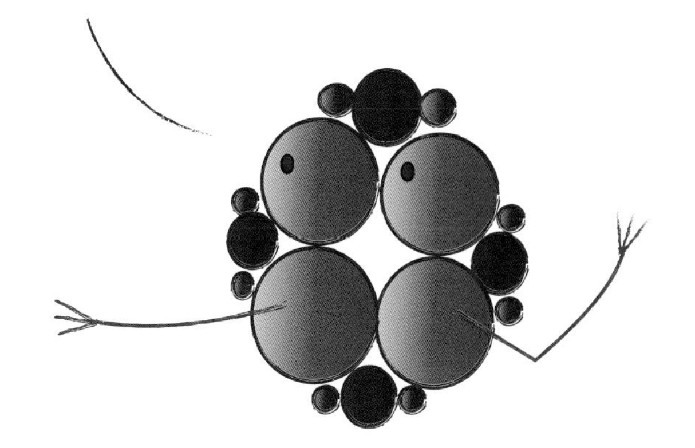

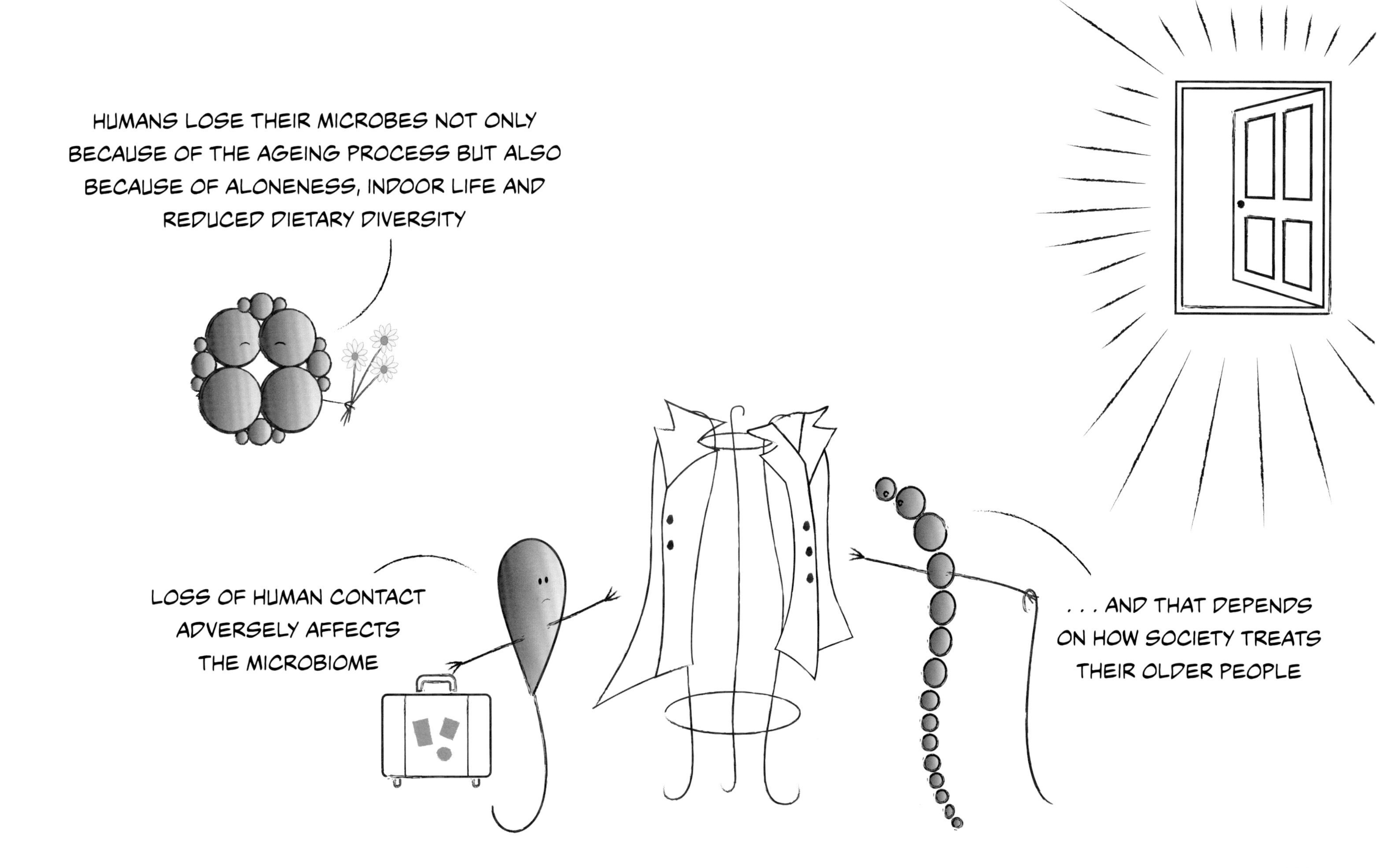
HUMANS LOSE THEIR MICROBES NOT ONLY BECAUSE OF THE AGEING PROCESS BUT ALSO BECAUSE OF ALONENESS, INDOOR LIFE AND REDUCED DIETARY DIVERSITY
LOSS OF HUMAN CONTACT ADVERSELY AFFECTS THE MICROBIOME
. . . AND THAT DEPENDS ON HOW SOCIETY TREATS THEIR OLDER PEOPLE

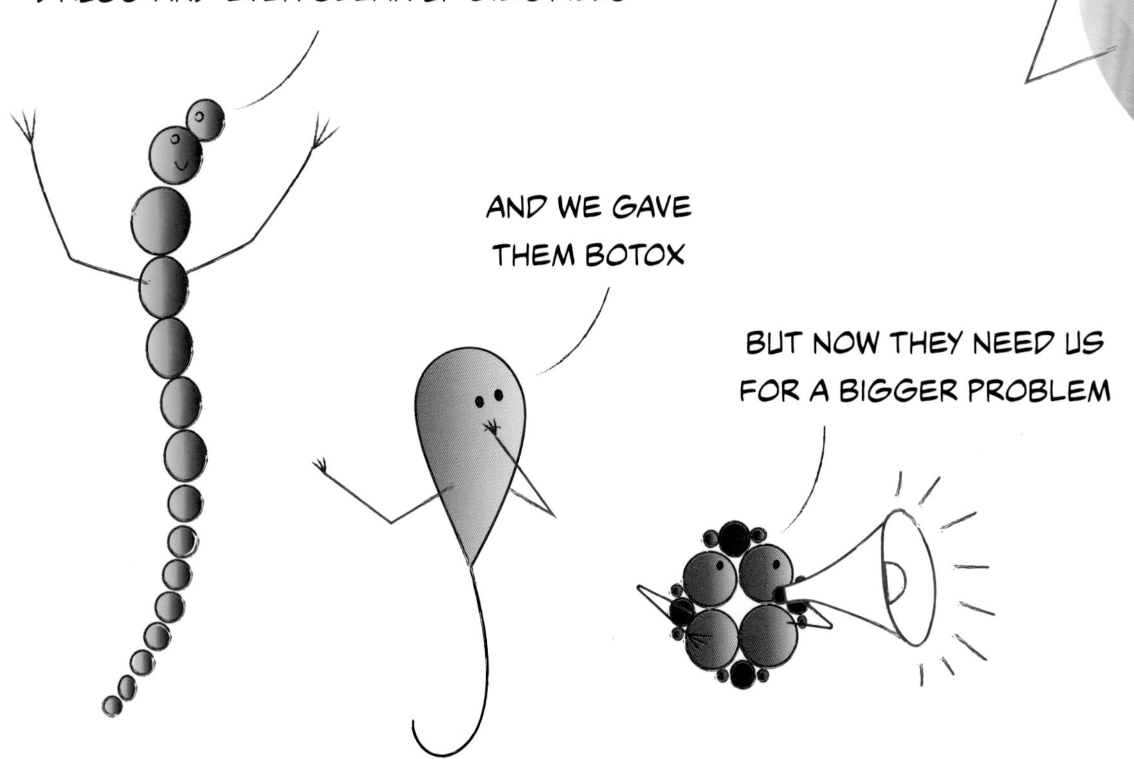
WE HAVE HELPED HUMANS TO BAKE BREAD, MAKE CHEESE, BREW BEER, PRODUCE LIFE-SAVING DRUGS AND EVEN CLEAN UP OIL SPILLS
AND WE GAVE THEM BOTOX
BUT NOW THEY NEED US FOR A BIGGER PROBLEM
HELP!

5 surviving the future

We can't predict the future, but we can project from current trends and from what happened in the past. Human activity has jeopardised the health of the planet. Now human survival is uncertain. The only certainty is that microbes will survive. Since microbes once made the planet habitable for humans, perhaps the solution for human survival lies with microbes. However, this will require better care and collaboration with nature.

In the words of the poet William Carlos Williams (1883-1963), 'Man has survived hitherto because he was too ignorant to know how to realise his wishes. Now that he can realise them, he must either change them or perish.'

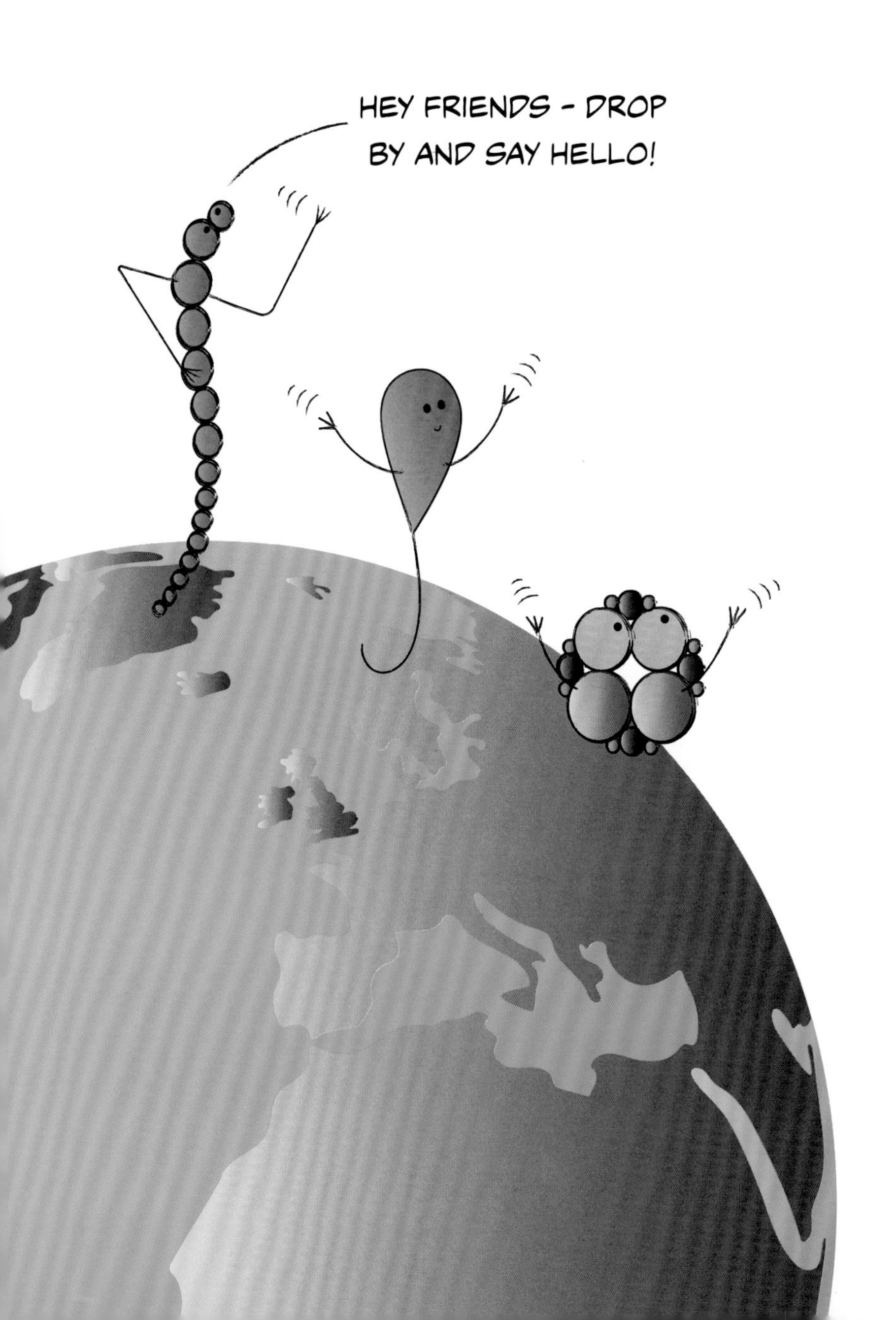
HEY FRIENDS - DROP
BY AND SAY HELLO!

THANKS BUT
NO THANKS!
YOU MICROBES MADE EARTH HABITABLE
BUT THE HUMANS ARE DESTROYING IT!

global warming and climate change

The most serious challenge facing humanity is caused by human activities which increase the concentration of greenhouse gases in the atmosphere. Greenhouse gases such as carbon dioxide, methane and nitrous oxide trap heat in the atmosphere. Microbes are an important part of the story. Since microbes produce and consume greenhouse gases, there is the prospect of changing human activity to favour microbial consumption rather than production of these gases.

Microbes are the unseen majority of creatures living on the planet. They are indicators of the health of the planet, and humans can learn a lot by studying how microbes will cope with global warming. Extreme climatic conditions that result from global warming will threaten the capacity of soil and oceanic microbes to sustain the food supply, whereas the distribution of pathogenic microbes may be altered, and increase the risk of infectious diseases in humans.

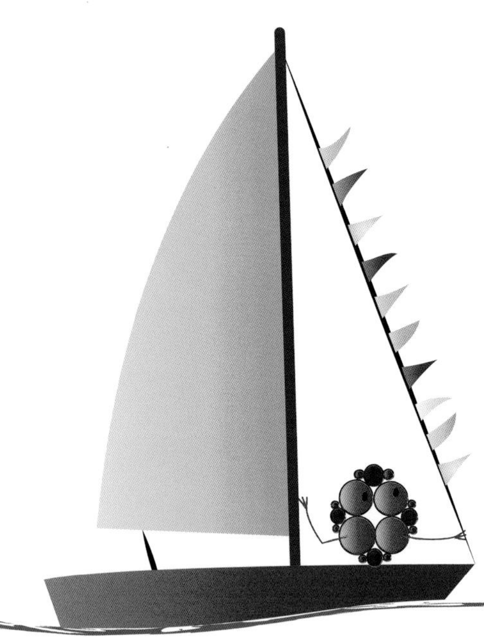

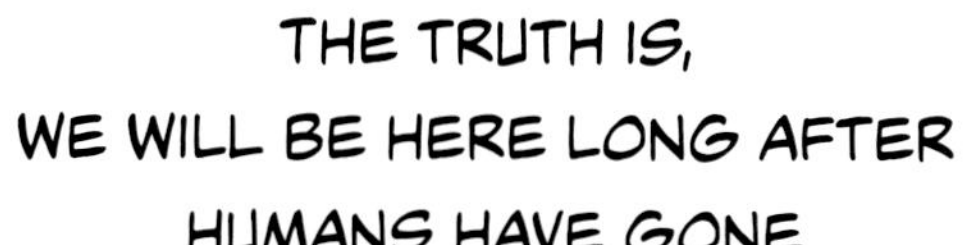
THE TRUTH IS,
WE WILL BE HERE LONG AFTER
HUMANS HAVE GONE

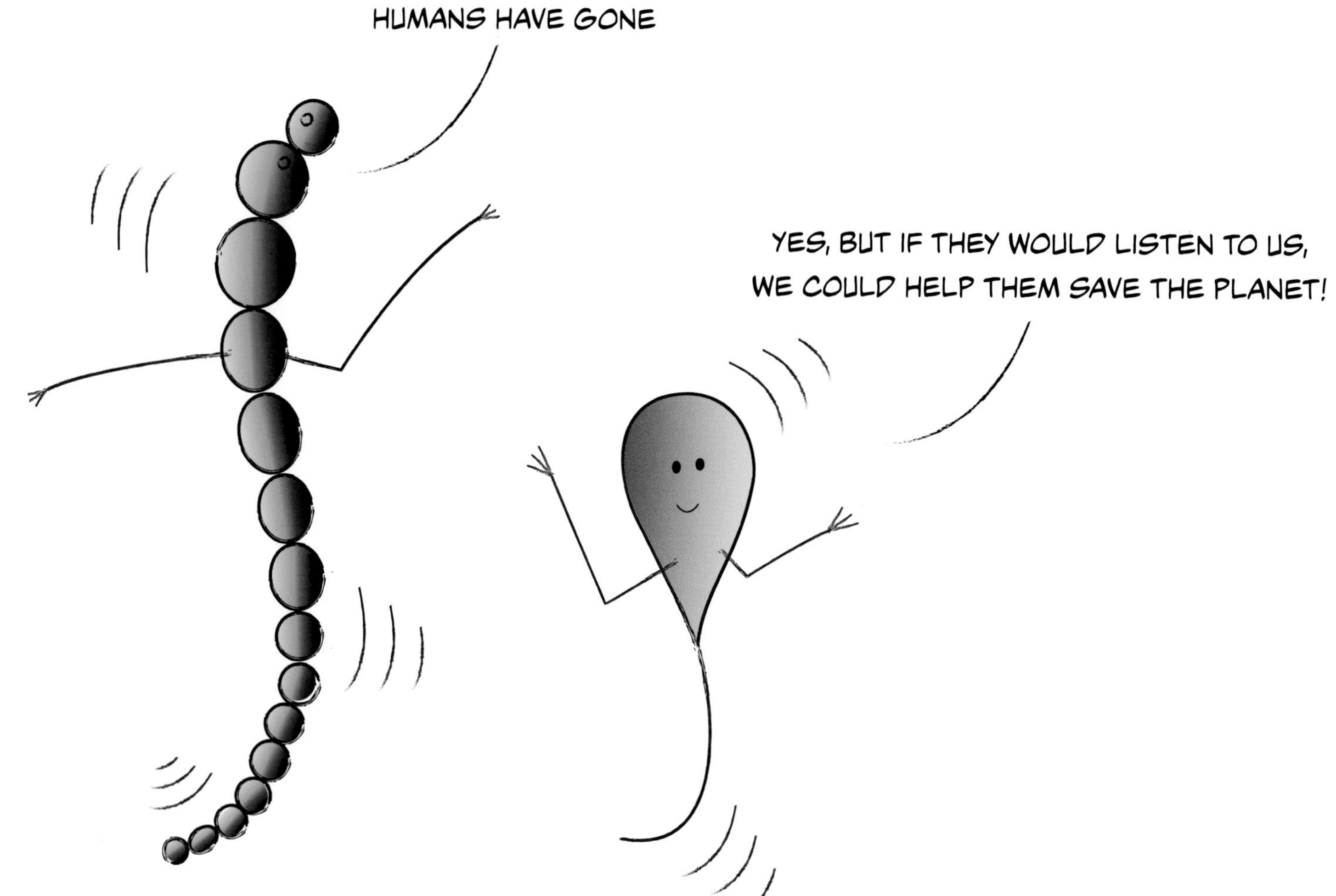
YES, BUT IF THEY WOULD LISTEN TO US,
WE COULD HELP THEM SAVE THE PLANET!

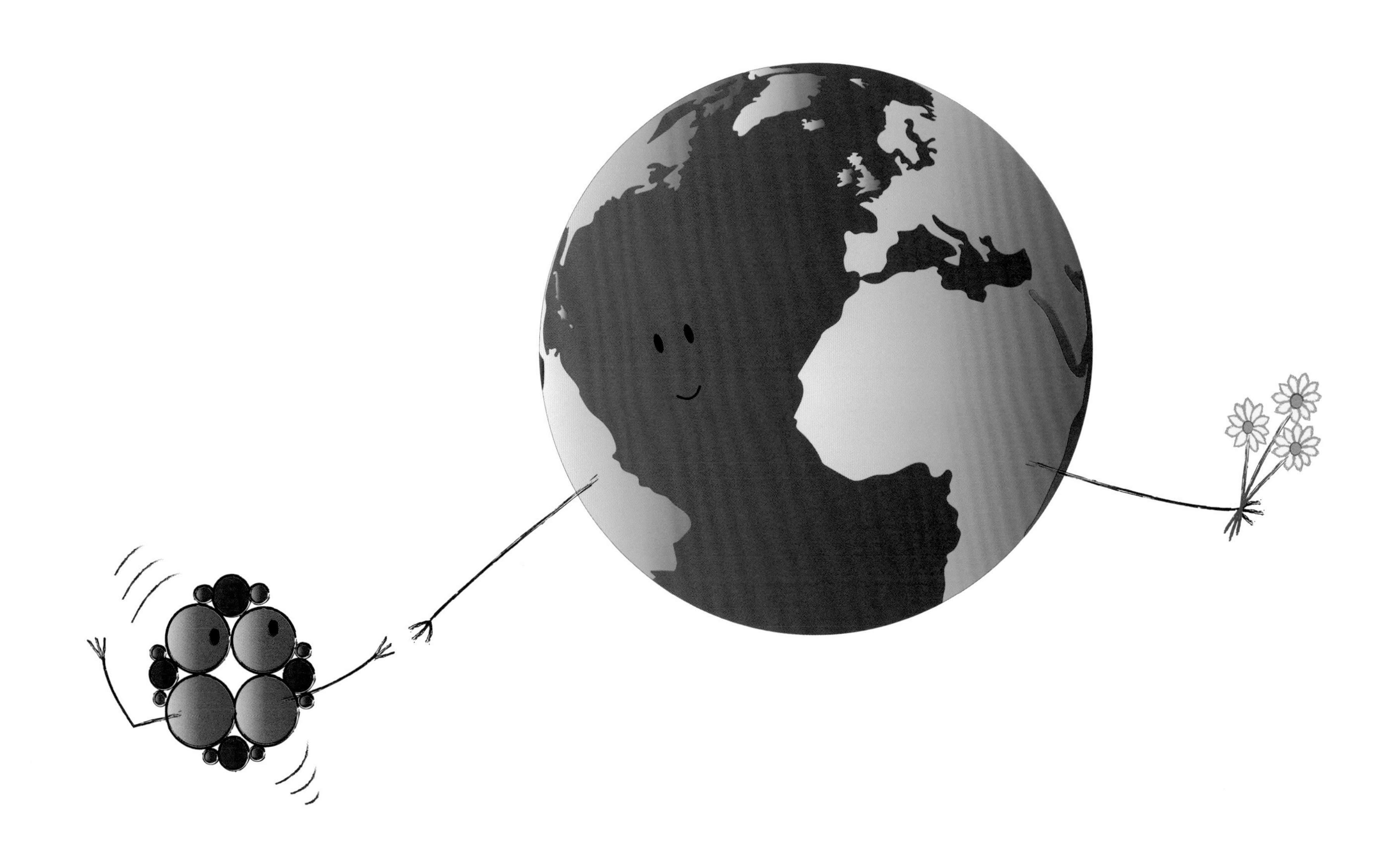

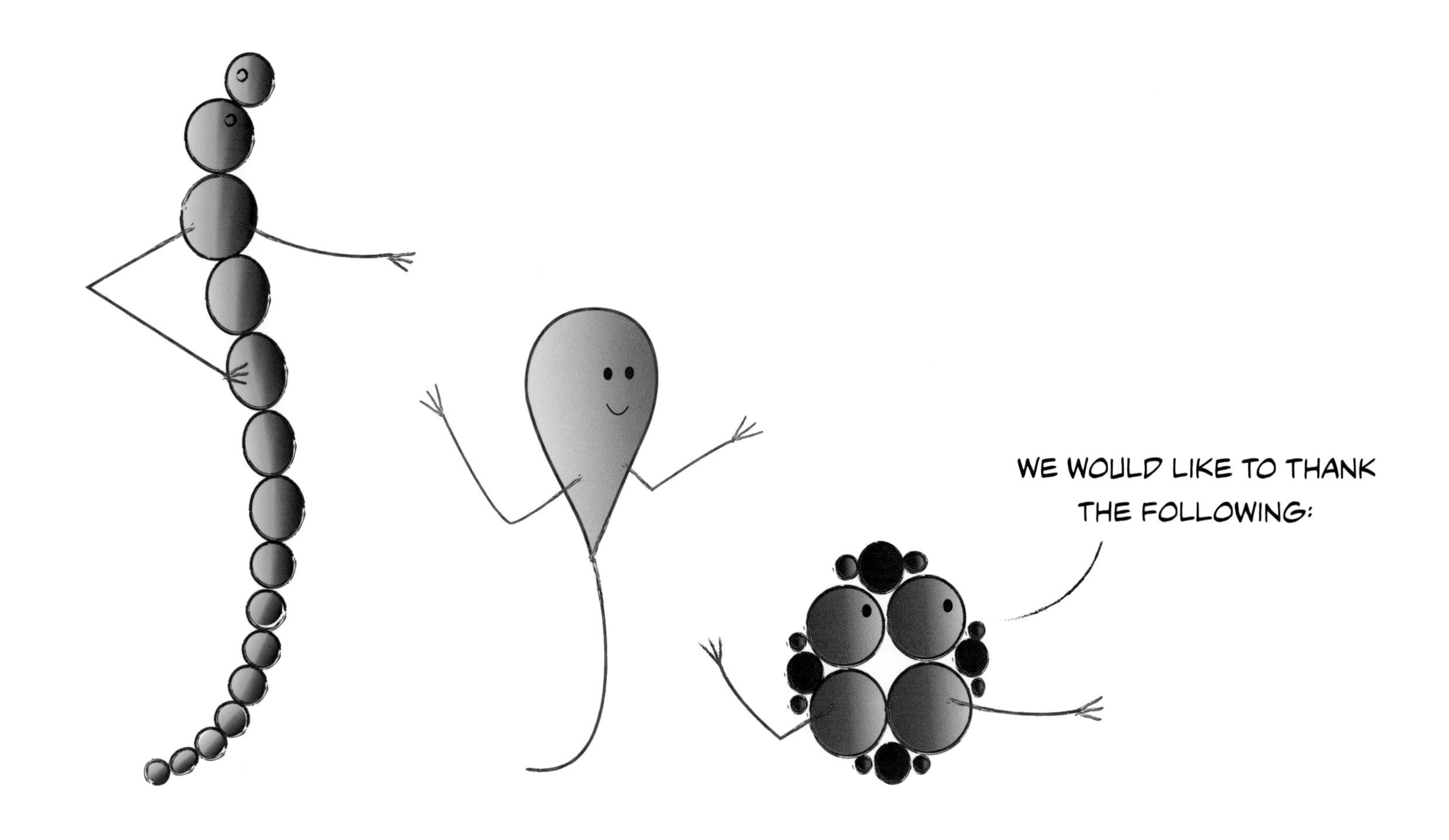
WE WOULD LIKE TO THANK
THE FOLLOWING:

acknowledgements

In support of public health education, the input of APC Microbiome Ireland, a Science Foundation Ireland-funded research centre, is acknowledged.

The kind sponsorship in the form of a philanthropic donation is particularly appreciated.

Special thanks to Professor Michael Molloy and to Bruce Francis. We also thank Seán O'Keeffe of Liberties Press for his guidance and the following for their input: Stephen Bean, Caoimhe Byrne, Sally Cudmore, Phil Gowers, Léna Guénebaut, Rosemary Shanahan, Lily Vass and Profs Gerard Fitzgerald, Paul O'Toole, Paul Ross and Joy Watts.

The views expressed are those of the authors and we alone are responsible for any errors or omissions.

fergus shanahan

Fergus Shanahan, photographed here with a replica of van Leeuwenhoek's microscope, is a clinician-scientist, a renowned gastroenterologist with more than forty years' experience helping people with chronic inflammatory bowel disease. He founded one of the world's first microbiome research centres – Science Foundation Ireland's APC Microbiome Ireland, and has received numerous international awards for contributions to medical science and the medical humanities. He loves microbes, words and people.

(photo: Stephen H. Bean)

laura gowers

Laura Gowers trained as an industrial design engineer in London. She started her career as a product designer, and for a short time lived and worked in the Netherlands, not far from where Antonie van Leeuwenhoek would have first seen these little animalcules! Interested in her own health, she became intrigued by the ingenuity of microbes, particularly the way in which microbes regulate the immune system in health and disease. In more recent years, her focus has moved towards graphic design and in particular the visual communication of complex scientific information in the form of accessible health messages to broad audiences.

(photo: Phil Gowers)

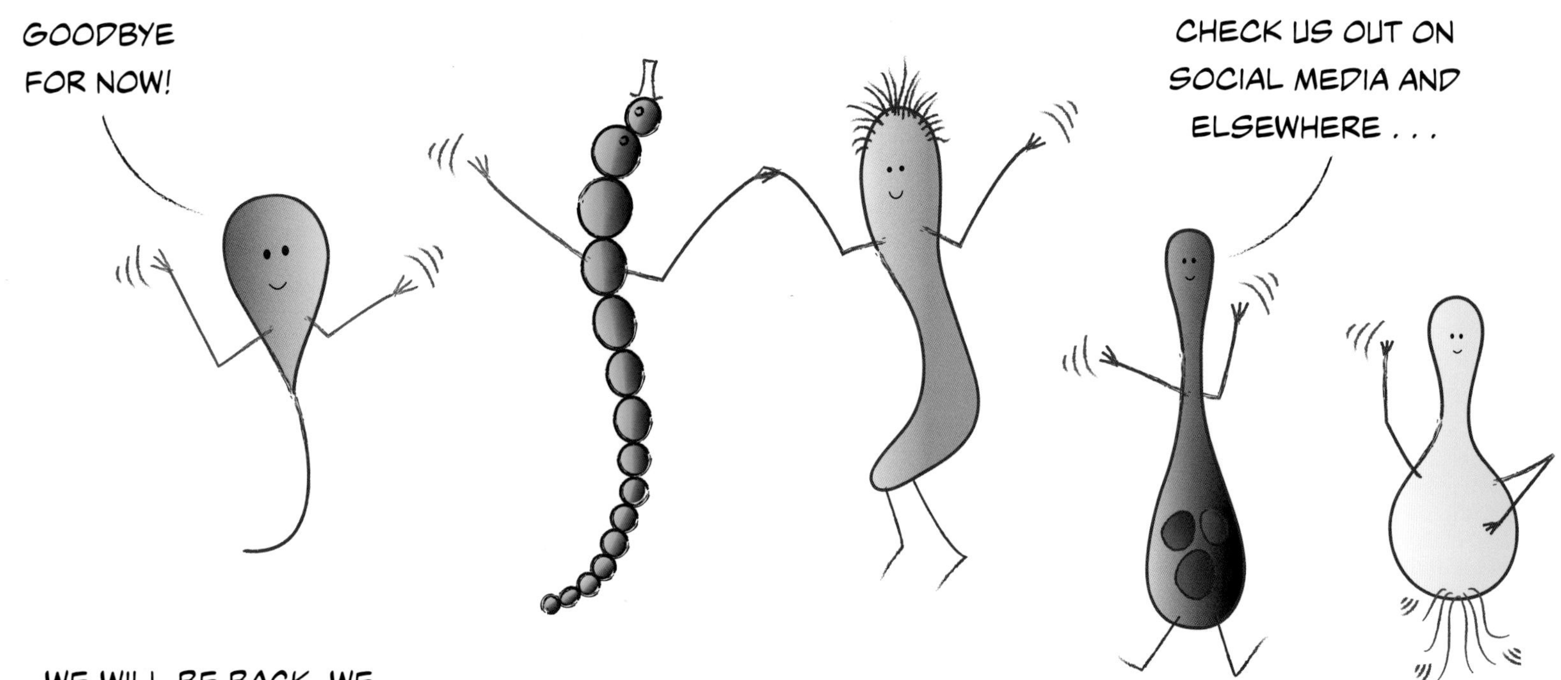

Use QR codes for further information on the microbiome and for an exclusive interview about how Fergus and Laura created the book *Listen to Your Microbes*

listentoyourmicrobes@gmail.com

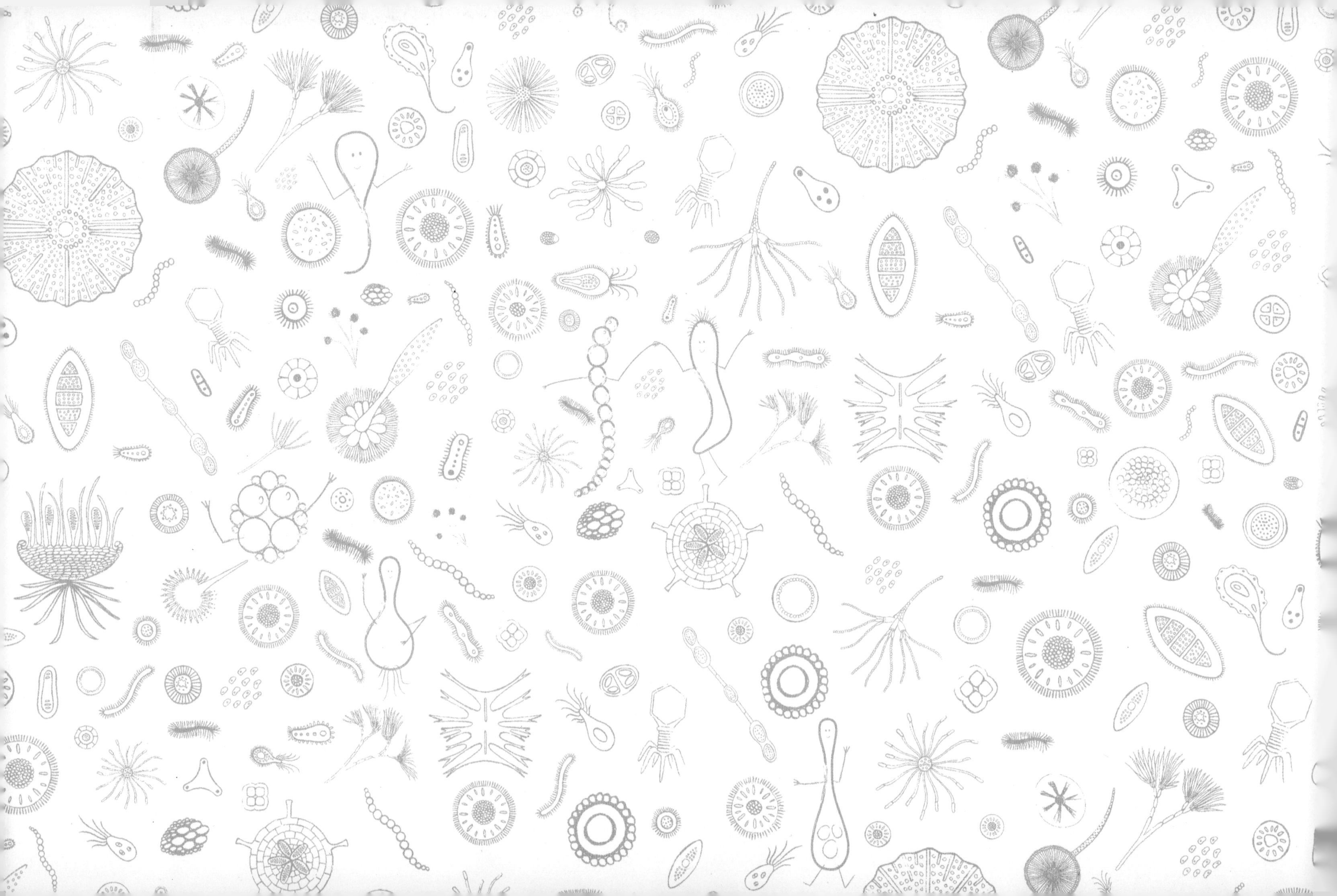